# Modern Astrophysics

# Modern Astrophysics

Cynthia Jefferson

Larsen & Keller
www.larsen-keller.com

Modern Astrophysics
Cynthia Jefferson
ISBN: 978-1-64172-401-2 (Hardback)

 Larsen & Keller

Published by Larsen and Keller Education,
5 Penn Plaza,
19th Floor,
New York, NY 10001, USA

**Cataloging-in-Publication Data**

Modern astrophysics / Cynthia Jefferson.
    p. cm.
Includes bibliographical references and index.
ISBN 978-1-64172-401-2
1. Astrophysics. 2 .Astronomy. 3. Cosmic physics.
4. Physics. I. Jefferson, Cynthia.
QB461 .M63 2020
523.01--dc23

For more information regarding Larsen and Keller Education and its products, please visit the publisher's website www.larsen-keller.com

# TABLE OF CONTENTS

| | | |
|---|---|---:|
| **Preface** | | **VII** |
| **Chapter 1** | **Astrophysics: An Introduction** | **1** |
| | • Astronomy | 1 |
| | • Astrophysics | 39 |
| **Chapter 2** | **Branches of Astrophysics** | **42** |
| | • Computational Astrophysics | 42 |
| | • Astroparticle Physics | 47 |
| | • Nuclear Astrophysics | 50 |
| | • Atomic and Molecular Astrophysics | 53 |
| | • Solar Physics | 54 |
| | • Astrophysical Fluid Dynamics | 57 |
| | • Stellar Dynamics | 61 |
| **Chapter 3** | **Basic Concepts in Astrophysics** | **65** |
| | • Astronomical Coordinate Systems | 65 |
| | • Astronomical Unit | 75 |
| | • Astronomical Unit of Mass | 81 |
| | • Big Bang Theory | 91 |
| | • Cosmic Microwave Background | 94 |
| | • Luminosity | 110 |
| | • Dark Matter | 116 |
| | • Dark Energy | 132 |
| | • Accretion | 134 |
| | • Optical Depth | 141 |

| Chapter 4 Astronomical Objects | 144 |
|---|---|
| • Celestial Bodies | 144 |
| • Star | 146 |
| • Galaxy | 179 |
| • Nebula | 181 |
| • Interstellar Medium | 186 |

| Chapter 5 Black Hole | 198 |
|---|---|
| • Event Horizon | 199 |
| • Gravitational Singularity | 200 |
| • Photon Sphere | 203 |
| • Ergosphere | 206 |
| • Hawking Radiation | 208 |

**Permissions**

**Index**

# PREFACE

This book has been written, keeping in view that students want more practical information. Thus, my aim has been to make it as comprehensive as possible for the readers. I would like to extend my thanks to my family and co-workers for their knowledge, support and encouragement all along.

The astronomical branch that is concerned with the application of principles of physics and chemistry to discover the nature of astronomical objects is referred to as astrophysics. Its central focus is on the study of celestial objects such as the sun, galaxies, the interstellar medium, extrasolar planets, and cosmic microwave background. Discharges from these objects are observed across the entire electromagnetic spectrum. Their properties such as density, temperature, chemical composition and luminosity are also studied in astrophysics. It draws on the concepts of various other disciplines including classical mechanics, electromagnetism, thermodynamics, quantum mechanics, relativity, nuclear and particle physics, as well as atomic and molecular physics. Some of the major branches of this field are observational and theoretical astrophysics. It also attempts to determine the properties of dark matter, black holes, dark energy and other celestial bodies. This textbook is a valuable compilation of topics, ranging from the basic to the most complex theories and principles in the field of astrophysics. Different approaches, evaluations and methodologies in this discipline have been included in this textbook. It will provide comprehensive knowledge to the readers.

A brief description of the chapters is provided below for further understanding:

Chapter – Astrophysics: An Introduction

The branch of astronomy that utilizes the principles of physics and chemistry to understand the nature of astronomical objects, such as the sun, stars and galaxies, is referred to as astrophysics. This is an introductory chapter which will introduce briefly all the significant aspects of astronomy as well as astrophysics.

Chapter – Branches of Astrophysics

Some of the various branches of astrophysics include solar physics, astroparticle physics, atomic and molecular astrophysics, computational astrophysics, nuclear astrophysics, astrophysical fluid dynamics and stellar dynamics. This chapter has been carefully written to provide an easy understanding of these branches of astrophysics.

Chapter – Basic Concepts in Astrophysics

Some of the fundamental concepts within the field of astrophysics are Big Bang theory, dark matter, dark energy, optical depth, accretion, cosmic microwave background and astronomical coordinate systems. The topics elaborated in this chapter such as will help in gaining a better perspective about these concepts of astrophysics.

Chapter – Astronomical Objects

The naturally occurring physical entity or structure that exists in the observable universe is known as an astronomical object. It includes celestial objects, stars, galaxies, nebulas, star clusters, etc. This chapter has been carefully written to provide an easy understanding of the various aspects of these astronomical objects.

Chapter – Black Hole

Black hole is a region of space-time that has extreme gravitational acceleration due to which no particles or electromagnetic radiation can escape from it. The chapter closely examines the properties and structure of black hole such as event horizon, ergosphere and photon sphere to provide an extensive understanding of the subject.

**Cynthia Jefferson**

# Astrophysics: An Introduction

**1**

- **Astronomy**
- **Astrophysics**

The branch of astronomy that utilizes the principles of physics and chemistry to understand the nature of astronomical objects, such as the sun, stars and galaxies, is referred to as astrophysics. This is an introductory chapter which will introduce briefly all the significant aspects of astronomy as well as astrophysics.

## Astronomy

Astronomy is the science that encompasses the study of all extraterrestrial objects and phenomena. Until the invention of the telescope and the discovery of the laws of motion and gravity in the 17th century, astronomy was primarily concerned with noting and predicting the positions of the Sun, Moon, and planets, originally for calendrical and astrological purposes and later for navigational uses and scientific interest. The catalog of objects now studied is much broader and includes, in order of increasing distance, the solar system, the stars that make up the Milky Way Galaxy, and other, more distant galaxies. With the advent of scientific space probes, Earth also has come to be studied as one of the planets, though its more-detailed investigation remains the domain of the Earth sciences.

### The Scope of Astronomy

Since the late 19th century astronomy has expanded to include astrophysics, the application of physical and chemical knowledge to an understanding of the nature of celestial objects and the physical processes that control their formation, evolution, and emission of radiation. In addition, the gases and dust particles around and between the stars have become the subjects of much research. Study of the nuclear reactions that provide the energy radiated by stars has shown how the diversity of atoms found in nature can be derived from a universe that, following the first few minutes of its existence, consisted only of hydrogen, helium, and a trace of lithium. Concerned with phenomena on the largest scale is cosmology, the study of the evolution of the universe.

Astrophysics has transformed cosmology from a purely speculative activity to a modern science capable of predictions that can be tested.

Its great advances notwithstanding, astronomy is still subject to a major constraint: it is inherently an observational rather than an experimental science. Almost all measurements must be performed at great distances from the objects of interest, with no control over such quantities as their temperature, pressure, or chemical composition. There are a few exceptions to this limitation—namely, meteorites (most of which are from the asteroid belt, though some are from the Moon or Mars), rock and soil samples brought back from the Moon, samples of comet and asteroid dust returned by robotic spacecraft, and interplanetary dust particles collected in or above the stratosphere. These can be examined with laboratory techniques to provide information that cannot be obtained in any other way. In the future, space missions may return surface materials from Mars, or other objects, but much of astronomy appears otherwise confined to Earth-based observations augmented by observations from orbiting satellites and long-range space probes and supplemented by theory.

Nickel-iron meteorite, from Canyon Diablo, Arizona.

## Determining Astronomical Distances

A central undertaking in astronomy is the determination of distances. Without a knowledge of astronomical distances, the size of an observed object in space would remain nothing more than an angular diameter and the brightness of a star could not be converted into its true radiated power, or luminosity. Astronomical distance measurement began with a knowledge of Earth's diameter, which provided a base for triangulation. Within the inner solar system, some distances can now be better determined through the timing of radar reflections or, in the case of the Moon, through laser ranging. For the outer planets, triangulation is still used. Beyond the solar system, distances to the closest stars are determined through triangulation, in which the diameter of Earth's orbit serves as the baseline and shifts in stellar parallax are the measured quantities. Stellar distances are commonly expressed by astronomers in parsecs (pc), kiloparsecs, or megaparsecs. (1 pc = $3.086 \times 10^{18}$ cm, or about 3.26 light-years [$1.92 \times 10^{13}$ miles].) Distances can be measured out to around a kiloparsec by trigonometric parallax (Determining stellar distances). The accuracy of measurements made from Earth's surface is limited by atmospheric effects, but measurements made from the Hipparcos satellite in the 1990s extended the scale to stars as far as 650 parsecs, with an accuracy of about a thousandth of an arc second. The Gaia satellite is expected to measure stars as far away as 10 kiloparsecs to an accuracy of 20 percent. Less-direct measurements must be used for more-distant stars and for galaxies.

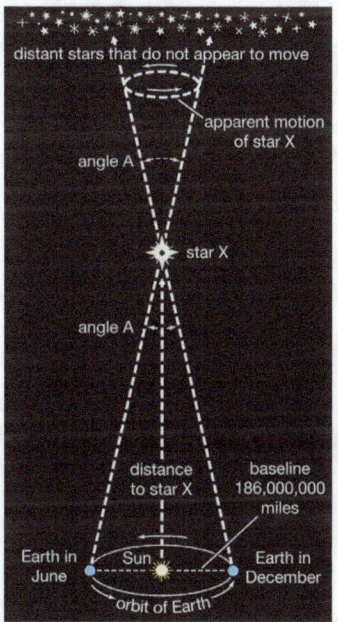

Calculating stellar distances.

Two general methods for determining galactic distances are described here. In the first, a clearly identifiable type of star is used as a reference standard because its luminosity has been well determined. This requires observation of such stars that are close enough to Earth that their distances and luminosities have been reliably measured. Such a star is termed a "standard candle." Examples are Cepheid variables, whose brightness varies periodically in well-documented ways, and certain types of supernova explosions that have enormous brilliance and can thus be seen out to very great distances. Once the luminosities of such nearer standard candles have been calibrated, the distance to a farther standard candle can be calculated from its calibrated luminosity and its actual measured intensity. (The measured intensity [I] is related to the luminosity [L] and distance [d] by the formula $I = L/4\pi d^2$.) A standard candle can be identified by means of its spectrum or the pattern of regular variations in brightness. (Corrections may have to be made for the absorption of starlight by interstellar gas and dust over great distances.) This method forms the basis of measurements of distances to the closest galaxies.

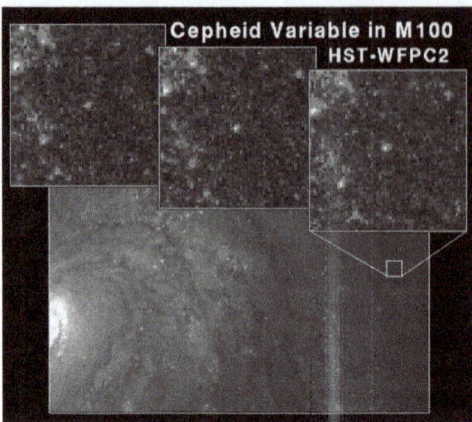

A region of the spiral galaxy M100 (bottom), with three frames (top) showing a Cepheid variable increasing in brightness. These images were taken with the Wide Field Planetary Camera 2 (WFPC2) on board the Hubble Space Telescope (HST).

The second method for galactic distance measurements makes use of the observation that the distances to galaxies generally correlate with the speeds with which those galaxies are receding from Earth (as determined from the Doppler shift in the wavelengths of their emitted light). This correlation is expressed in the Hubble law: velocity = H × distance, in which H denotes Hubble's constant, which must be determined from observations of the rate at which the galaxies are receding. There is widespread agreement that H lies between 67 and 73 kilometres per second per megaparsec (km/sec/Mpc). H has been used to determine distances to remote galaxies in which standard candles have not been found.

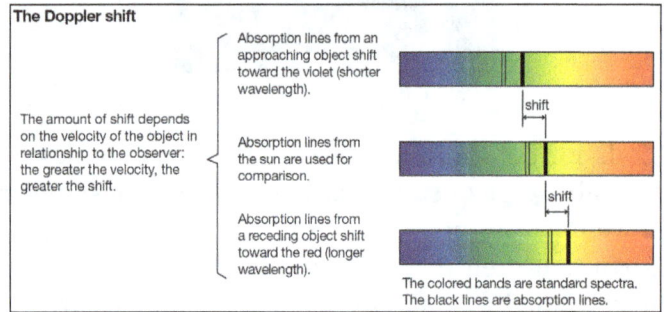

Doppler shift.

## Study of the Solar System

The solar system took shape 4.57 billion years ago, when it condensed within a large cloud of gas and dust. Gravitational attraction holds the planets in their elliptical orbits around the Sun. In addition to Earth, five major planets (Mercury, Venus, Mars, Jupiter, and Saturn) have been known from ancient times. Since then only two more have been discovered: Uranus by accident in 1781 and Neptune in 1846 after a deliberate search following a theoretical prediction based on observed irregularities in the orbit of Uranus. Pluto, discovered in 1930 after a search for a planet predicted to lie beyond

Neptune, was considered a major planet until 2006, when it was redesignated a dwarf planet by the International Astronomical Union.

Jupiter, the fifth planet from the Sun and largest planet in the solar system. The Great Red Spot is visible in the lower left. This image is based on observations made by the Voyager 1 spacecraft in 1979.

The average Earth-Sun distance, which originally defined the astronomical unit (AU), provides a convenient measure for distances within the solar system. The astronomical unit was originally defined by observations of the mean radius of Earth's orbit but is now defined as 149,597,870.7 km (about 93 million miles). Mercury, at 0.4 AU, is the closest planet to the Sun, while Neptune, at 30.1 AU, is the farthest. Pluto's orbit, with a mean radius of 39.5 AU, is sufficiently eccentric that at times it is closer to the Sun than is Neptune. The planes of the planetary orbits are all within a few degrees of the ecliptic, the plane that contains Earth's orbit around the Sun. As viewed from far above Earth's North Pole, all planets move in the same (counterclockwise) direction in their orbits.

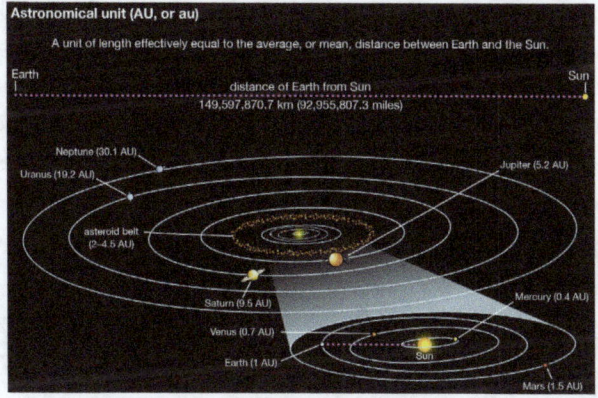

Scale of the solar system.

Distances in the solar system as measured in astronomical units (AU). Most of the mass of the solar system is concentrated in the Sun, with its $1.99 \times 10^{33}$ grams. Together, all of the planets amount to $2.7 \times 10^{30}$ grams (i.e., about one-thousandth of the Sun's mass), and Jupiter alone accounts for 71 percent of this amount. The solar system also contains five known objects of intermediate size classified as dwarf planets and a very large number of much smaller objects collectively called small bodies. The small bodies, roughly in order of decreasing size, are the asteroids, or minor planets; comets, including Kuiper belt, Centaur, and Oort cloud objects; meteoroids; and interplanetary dust particles. Because of their starlike appearance when discovered, the largest of these bodies were termed *asteroids*, and that name is widely used, but, now that the rocky nature of these bodies is understood, their more descriptive name is minor planets.

The eight planets of the solar system and Pluto, in a montage of images scaled to show the approximate sizes of the bodies relative to one another. Outward from the Sun, which is represented to scale by the yellow segment at the extreme left, are the four rocky terrestrial planets (Mercury, Venus, Earth, and Mars), the four hydrogen-rich giant planets (Jupiter, Saturn, Uranus, and Neptune), and icy, comparatively tiny Pluto.

The four inner, terrestrial planets—Mercury, Venus, Earth, and Mars—along with the Moon have average densities in the range of 3.9–5.5 grams per cubic cm, setting them apart from the four outer, giant planets—Jupiter, Saturn, Uranus, and Neptune—whose densities are all close to 1 gram per cubic cm, the density of water. The compositions of these two groups of planets must therefore be significantly different. This dissimilarity is thought to be attributable to conditions that prevailed during the early development of the solar system. Planetary temperatures now range from around 170 °C (330 °F, 440 K) on Mercury's surface through the typical 15 °C (60 °F, 290 K) on Earth to −135 °C (−210 °F, 140 K) on Jupiter near its cloud tops and down to −210 °C (−350 °F, 60 K) near Neptune's cloud tops. These are average temperatures; large variations exist between dayside and nightside for planets closest to the Sun, except for Venus with its thick atmosphere.

Saturn and its spectacular rings, in a natural-colour composite of 126 images taken by the Cassini spacecraft on October 6, 2004. The view is directed toward Saturn's southern hemisphere, which is tipped toward the Sun. Shadows cast by the rings are visible against the bluish northern hemisphere, while the planet's shadow is projected on the rings to the left.

The surfaces of the terrestrial planets and many satellites show extensive cratering, produced by high-speed impacts. On Earth, with its large quantities of water and an active atmosphere, many of these cosmic footprints have eroded, but remnants of very large craters can be seen in aerial and spacecraft photographs of the terrestrial surface. On Mercury, Mars, and the Moon, the absence of water and any significant atmosphere has left the craters unchanged for billions of years, apart from disturbances produced by infrequent later impacts. Volcanic activity has been an important force in the shaping of the surfaces of the Moon and the terrestrial planets. Seismic activity on the Moon has been monitored by means of seismometers left on its surface by Apollo astronauts and by Lunokhod robotic rovers. Cratering on the largest scale seems to have ceased about three billion years ago, although on the Moon there is clear evidence for a continued cosmic drizzle of small particles, with the larger objects churning ("gardening") the lunar surface and the smallest producing microscopic impact pits in crystals in the lunar rocks.

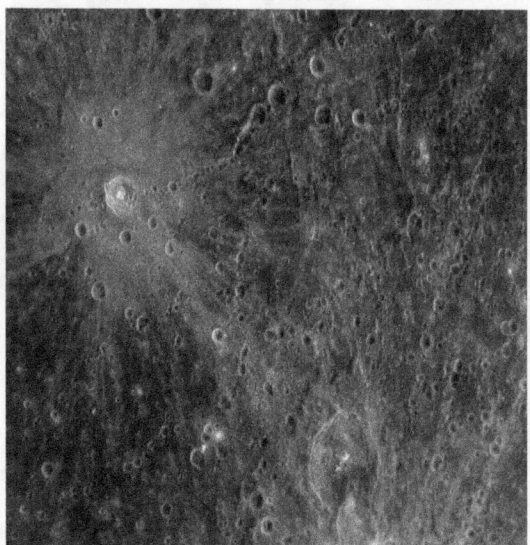

Mercury: Meteorite crater surrounded by rays of ejected material on Mercury, in a photograph taken by the Messenger probe, January 2008. A chain of craters crosses the centre of the rayed crater.

All of the planets apart from the two closest to the Sun (Mercury and Venus) have natural satellites (moons) that are very diverse in appearance, size, and structure, as revealed in close-up observations from long-range space probes. The four outer dwarf planets have moons; Pluto has at least five moons, including one, Charon, fully half the size of Pluto itself. Over 200 asteroids and 80 Kuiper belt objects also have moons. Four planets (Jupiter, Saturn, Uranus, and Neptune), one dwarf planet (Haumea), and one Centaur object (Chariklo) have rings, disklike systems of small rocks and particles that orbit their parent bodies.

Pluto's moon Charon.

## Lunar Exploration

During the U.S. Apollo missions a total weight of 381.7 kg (841.5 pounds) of lunar material was collected; an additional 300 grams (0.66 pounds) was brought back by unmanned Soviet Luna vehicles. About 15 percent of the Apollo samples have been distributed for analysis, with the remainder stored at the NASA Johnson Space Center, Houston, Texas. The opportunity to employ a wide range of laboratory techniques on these lunar samples has revolutionized planetary science. The results of the analyses have enabled investigators to determine the composition and age of the lunar surface. Seismic observations have made it possible to probe the lunar interior. In addition, retroreflectors left on the Moon's surface by Apollo astronauts have allowed high-power laser beams to be sent from Earth to the Moon and back, permitting scientists to monitor the Earth-Moon distance to an accuracy of a few centimetres. This experiment, which has provided data used in calculations of the dynamics of the Earth-Moon system, has shown that the separation of the two bodies is increasing by 4.4 cm (1.7 inches) each year.

Apollo 17 geologist-astronaut Harrison Schmitt at the foot of a huge split boulder, December 1972, during the mission's third extravehicular exploration of the Taurus-Littrow Valley landing site.

## Planetary Studies

Mercury is too hot to retain an atmosphere, but Venus's brilliant white appearance is the result of its being completely enveloped in thick clouds of carbon dioxide, impenetrable at visible wavelengths. Below the upper clouds, Venus has a hostile atmosphere containing clouds of sulfuric acid droplets. The cloud cover shields the planet's surface from direct sunlight, but the energy that does filter through warms the surface, which then radiates at infrared wavelengths. The long-wavelength infrared radiation is trapped by the dense clouds such that an efficient greenhouse effect keeps the surface temperature near 465 °C (870 °F, 740 K). Radar, which can penetrate the thick Venusian clouds, has been used to map the planet's surface. In contrast, the atmosphere of Mars is very thin and is composed mostly of carbon dioxide (95 percent), with very little water vapour; the planet's surface pressure is only about 0.006 that of Earth. The outer planets have atmospheres composed largely of light gases, mainly hydrogen and helium.

Photo mosaic of Mercury, taken by the Mariner 10 spacecraft, 1974.

Each planet rotates on its axis, and nearly all of them rotate in the same direction—counterclockwise as viewed from above the ecliptic. The two exceptions are Venus, which rotates in the clockwise direction beneath its cloud cover, and Uranus, which has its rotation axis very nearly in the plane of the ecliptic.

Some of the planets have magnetic fields. Earth's field extends outward until it is disturbed by the solar wind—an outward flow of protons and electrons from the Sun—which carries a magnetic field along with it. Through processes not yet fully understood, particles from the solar wind and galactic cosmic rays (high-speed particles from outside the solar system) populate two doughnut-shaped regions called the Van Allen radiation belts. The inner belt extends from about 1,000 to 5,000 km (600 to 3,000 miles) above Earth's surface, and the outer from roughly 15,000 to 25,000 km (9,300 to 15,500 miles). In these belts, trapped particles spiral along paths that take them around Earth while bouncing back and forth between the Northern and Southern hemispheres, with their orbits controlled by Earth's magnetic field. During periods of increased solar activity, these regions of trapped particles are disturbed, and some of the particles move down into Earth's atmosphere, where they collide with atoms and molecules to produce auroras.

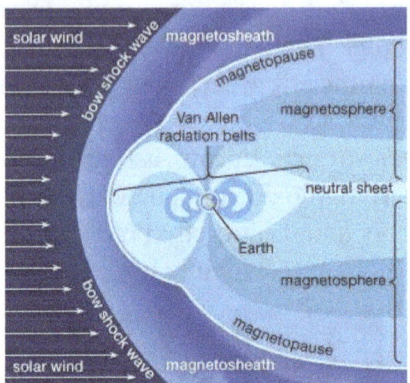

The Van Allen radiation belts contained within Earth's magnetosphere. Pressure from the solar wind is responsible for the asymmetrical shape of the magnetosphere and the belts.

Jupiter has a magnetic field far stronger than Earth's and many more trapped electrons, whose synchrotron radiation (electromagnetic radiation emitted by high-speed charged particles that are forced to move in curved paths, as under the influence of a magnetic field) is detectable from Earth. Bursts of increased radio emission are correlated with the position of Io, the innermost of the four Galilean moons of Jupiter. Saturn has a magnetic field that is much weaker than Jupiter's, but it too has a region of trapped particles. Mercury has a weak magnetic field that is only about 1 percent as strong as Earth's and shows no evidence of trapped particles. Uranus and Neptune have fields that are less than one-tenth the strength of Saturn's and appear much more complex than that of Earth. No field has been detected around Venus or Mars.

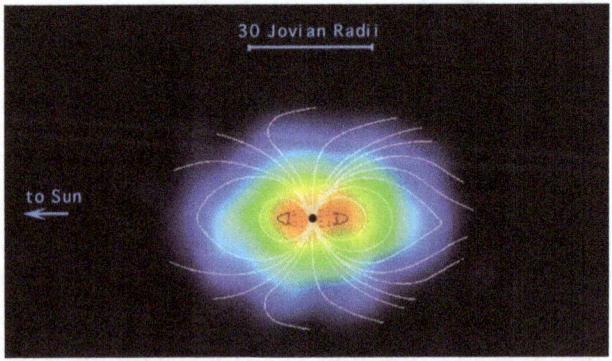

Jupiter's magnetosphere as observed by the Cassini spacecraft in 2000. The magnetosphere is the largest object in the solar system.

## Investigations of the Smaller Bodies

More than 500,000 asteroids with well-established orbits are known, and thousands of additional objects are discovered each year. Hundreds of thousands more have been seen, but their orbits have not been as well determined. It is estimated that several million asteroids exist, but most are small, and their combined mass is estimated to be less than a thousandth that of Earth. Most of the asteroids have orbits close to the ecliptic and move in the asteroid belt, between 2.3 and 3.3 AU from the Sun. Because some asteroids travel in orbits that can bring them close to Earth, there is a possibility of a collision that could have devastating results.

Asteroid Gaspra, composite of two images taken by the Galileo spacecraft on October 29, 1991. Galileo observed some 600 impact craters on Gaspra, from the large concavity visible on the lower right to craters as small as 100 metres (330 feet) in diameter. The asteroid's irregular shape and fracture lines suggest that it was once part of a larger body.

Comets are considered to come from a vast reservoir, the Oort cloud, orbiting the Sun at distances of 20,000–50,000 AU or more and containing trillions of icy objects—latent comet nuclei—with the potential to become active comets. Many comets have been observed over the centuries. Most make only a single pass through the inner solar system, but some are deflected by Jupiter or Saturn into orbits that allow them to return at predictable times. Halley's Comet is the best known of these periodic comets; its next return into the inner solar system is predicted for 2061. Many short-period comets are thought to come from the Kuiper belt, a region lying mainly between 30 AU and 50 AU from the Sun—beyond Neptune's orbit but including part of Pluto's—and housing perhaps hundreds of millions of comet nuclei. Very few comet masses have been well determined, but most are probably less than 1018 grams, one-billionth the mass of Earth.

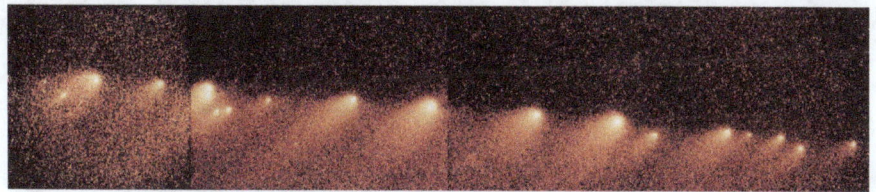

Composite image of the fragments of Comet Shoemaker-Levy 9.

Since the 1990s more than a thousand comet nuclei in the Kuiper belt have been observed with large telescopes; a few are about half the size of Pluto, and Pluto is the largest Kuiper belt object. Pluto's orbital and physical characteristics had long caused it to be regarded as an anomaly among the planets. However, after the discovery of numerous other Pluto-like objects beyond Neptune, Pluto was seen to be no longer unique in its "neighbourhood" but rather a giant member of the local population. Consequently, in 2006 astronomers at the general assembly of the International Astronomical Union elected to create the new category of dwarf planets for objects with such qualifications. Pluto, Eris, and Ceres, the latter being the largest member of the asteroid belt, were given this distinction. Two other Kuiper belt objects, Makemake and Haumea, were also designated as dwarf planets.

Pluto as seen by the New Horizons spacecraft.

Smaller than the observed asteroids and comets are the meteoroids, lumps of stony or metallic material believed to be mostly fragments of asteroids. Meteoroids vary from small rocks to boulders weighing a ton or more. A relative few have orbits that bring them into Earth's atmosphere and down to the surface as meteorites. Most meteorites that have been collected on Earth are probably from asteroids. A few have been identified as being from the Moon, Mars, or the asteroid Vesta.

Chelyabinsk meteorite: A cloud trail left behind by a meteorite that later exploded over Chelyabinsk province, Russia.

Meteorites are classified into three broad groups: stony (chondrites and achondrites; about 94 percent), iron (5 percent), and stony-iron (1 percent). Most meteoroids that enter the atmosphere heat up sufficiently to glow and appear as meteors, and the great majority of these vaporize completely or break up before they reach the surface. Many, perhaps most, meteors occur in showers and follow orbits that seem to be identical with those of certain comets, thus pointing to a cometary origin. For example, each May, when Earth crosses the orbit of Halley's Comet, the Eta Aquarid meteor shower occurs. Micrometeorites (interplanetary dust particles), the smallest meteoroidal particles, can be detected from Earth-orbiting satellites or collected by specially equipped aircraft flying in the stratosphere and returned for laboratory inspection. Since the late 1960s numerous meteorites have been found in the Antarctic on the surface of stranded ice flows. Some meteorites contain microscopic crystals whose isotopic proportions are unique and appear to be dust grains that formed in the atmospheres of different stars.

Hoba meteorite, lying where it was discovered in 1920 in Grootfontein, Namibia. The object, the largest meteorite known and an iron meteorite by classification, is made of nickel-iron alloy and estimated to weigh nearly 60 tons.

## Determinations of Age and Chemical Composition

The age of the solar system, taken to be close to 4.6 billion years, has been derived from measurements of radioactivity in meteorites, lunar samples, and Earth's crust. Abundances of isotopes of uranium, thorium, and rubidium and their decay products, lead and strontium, are the measured quantities.

Assessment of the chemical composition of the solar system is based on data from Earth, the Moon, and meteorites as well as on the spectral analysis of light from the Sun and planets. In broad outline, the solar system abundances of the chemical elements decrease with increasing atomic weight. Hydrogen atoms are by far the most abundant, constituting 91 percent; helium is next, with 8.9 percent; and all other types of atoms together amount to only 0.1 percent.

## Theories of Origin

The origin of Earth, the Moon, and the solar system as a whole is a problem that has not yet been settled in detail. The Sun probably formed by condensation of the central region of a large cloud of gas and dust, with the planets and other bodies of the solar system forming soon after, their composition strongly influenced by the temperature and pressure gradients in the evolving solar nebula. Less-volatile materials could condense into solids relatively close to the Sun to form the terrestrial planets. The abundant, volatile lighter elements could condense only at much greater distances to form the giant gas planets.

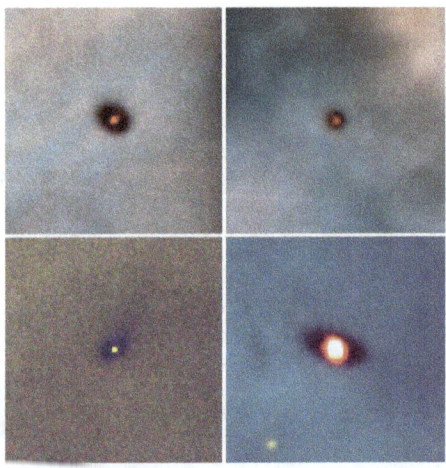

Four recently formed stars in the Orion nebula show what the solar system might have looked like at its birth some 4.6 billion years ago. Gravity collapsed dense globules of dust and gas into each of the four stars. The outer material formed into the surrounding protoplanetary disks, which are believed to be the precursors of planets. The bright glow in each image is the star, and the dark oval is the protoplanetary disk seen at an angle.

In the 1990s astronomers confirmed that other stars have one or more planets revolving around them. Studies of these planetary systems have both supported and challenged astronomers' theoretical models of how Earth's solar system formed. Unlike the solar system, many extrasolar planetary systems have large gas giants like Jupiter orbiting very close to their stars, and in some cases these "hot Jupiters" are closer to their star than Mercury is to the Sun.

That so many gas giants, which form in the outer regions of their system, end up so close to their stars suggests that gas giants migrate and that such migration may have happened in the solar system's history. According to the Grand Tack hypothesis, Jupiter may have done so within a few million years of the solar system's formation. In this scenario, Jupiter is the first giant planet to form, at about 3 AU from the Sun. Drag from the protoplanetary disk causes it to fall inward to about 1.5 AU. However, by this time, Saturn begins to form at about 3 AU and captures Jupiter in a 3:2 resonance. (That is, for every three revolutions Jupiter makes, Saturn makes two.) The two planets migrate outward and clear away any material that would have gone to making Mars bigger. Mars should be bigger than Venus or Earth, but it is only half their size. The Grand Tack, in which Jupiter moves inward and then outward, explains Mars's small size.

About 500 million years after the Grand Tack, according to the Nice Model (named after the French city where it was first proposed), after the four giant planets—Jupiter, Saturn, Uranus, and Neptune—formed, they orbited 5–17 AU from the Sun. These planets were in a disk of smaller bodies called planetesimals and in orbital resonances with each other. About four billion years ago, gravitational interactions with the planetesimals increased the eccentricity of the planets' orbits, driving them out of resonance. Saturn, Uranus and Neptune migrated outward, and Jupiter migrated slightly inward. (Uranus and Neptune may even have switched places.) This migration scattered the disk, causing the Late Heavy Bombardment. The final remnant of the disk became the Kuiper belt.

The origin of the planetary satellites is not entirely settled. As to the origin of the Moon, the opinion of astronomers long oscillated between theories that saw its origin and condensation as simultaneous with the formation of Earth and those that posited a separate origin for the Moon and its later capture by Earth's gravitational field. Similarities and differences in abundances of the chemical elements and their isotopes on Earth and the Moon challenged each group of theories. Finally, in the 1980s a model emerged that gained the support of most lunar scientists—that of a large impact on Earth and the expulsion of material that subsequently formed the Moon. For the outer planets, with their multiple satellites, many very small and quite unlike one another, the picture is less clear. Some of these moons have relatively smooth icy surfaces, whereas others are heavily cratered; at least one, Jupiter's Io, is volcanic. Some of the moons may have formed along with their parent planets, and others may have formed elsewhere and been captured.

## Study of Extrasolar Planetary Systems

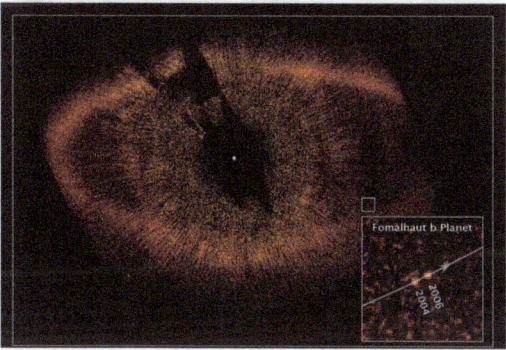

The extrasolar planet Fomalhaut b in images taken by the Hubble Space Telescope in 2004 and 2006. The black spot at the centre of the image is a coronagraph used to block the light from Fomalhaut, which is located at the white dot. The oval ring is Fomalhaut's dust belt, and the lines radiating from the centre of the image are scattered starlight.

The first extrasolar planets were discovered in 1992, and more than 3,700 such planets are known. Over 600 of these systems have more than one planet. Because planets are much fainter than their stars, fewer than 100 have been imaged directly. Most extrasolar planets have been found through their transit, the small dimming of a star's light when a planet passes in front of it.

Many of these planets are unlike those of the solar system. Hot Jupiters are large gas giants that orbit very close to their star. For example, HD 209458b is 0.69 times the mass of Jupiter and orbits its star every 3.52 days. Hot Neptunes are large ice giants about 10 percent of Jupiter's mass that also orbit very close to their star. Super-Earths are planets that are likely rocky like Earth but several times larger.

Artist's conception of the extrasolar planet HD 209458 b, some 150 light-years from Earth.

A primary goal of extrasolar planet research has been finding another planet that could support life. A useful guide for finding a life-supporting planet has been the concept of a habitable zone, the distance from a star where liquid water could survive on a planet's surface. About 40 planets have been found that are roughly Earth-sized and orbit in a habitable zone.

Artist's conception of Kepler-452b, the first approximately Earth-sized planet to be found in the habitable zone of a star like the Sun.

## Study of the Stars

## Measuring Observable Stellar Properties

The measurable quantities in stellar astrophysics include the externally observable features of the stars: distance, temperature, radiation spectrum and luminosity, composition (of the outer layers), diameter, mass, and variability in any of these. Theoretical astrophysicists use these observations to model the structure of stars and to devise theories for their formation and evolution. Positional information can be used for dynamical analysis, which yields estimates of stellar masses.

Bright nebulosity in the Pleiades (M45, NGC 1432), distance 490 light-years. Cluster stars provide the light, and surrounding clouds of dust reflect and scatter the rays from the stars.

In a system dating back at least to the Greek astronomer-mathematician Hipparchus in the 2nd century BCE, apparent stellar brightness ($m$) is measured in magnitudes. Magnitudes are now defined such that a first-magnitude star is 100 times brighter than a star of sixth magnitude. The human eye cannot see stars fainter than about sixth magnitude, but modern instruments used with large telescopes can record stars as faint as about 30th magnitude. By convention, the absolute magnitude ($M$) is defined as the magnitude that a star would appear to have if it were located at a standard distance of 10 parsecs. These quantities are related through the expression $m - M = 5 \log_{10} r - 5$, in which $r$ is the star's distance in parsecs.

The magnitude scale is anchored on a group of standard stars. An absolute measure of radiant power is luminosity, which is related to the absolute magnitude and usually expressed in ergs per second (ergs/sec). (Sometimes the luminosity is stated in terms of the solar luminosity, $3.86 \times 10^{33}$ ergs/sec.) Luminosity can be calculated when $m$ and $r$ are known. Correction might be necessary for the interstellar absorption of starlight.

There are several methods for measuring a star's diameter. From the brightness and distance, the luminosity ($L$) can be calculated, and, from observations of the brightness at different wavelengths, the temperature ($T$) can be calculated. Because the radiation from many stars can be well approximated by a Planck blackbody spectrum these measured quantities can be related through the expression $L = 4\pi R^2 \sigma T^4$, thus providing a means of calculating $R$, the star's radius. In this expression, $\sigma$ is the Stefan-Boltzmann constant, $5.67 \times 10^{-5}$ ergs/cm$^{2K4}$sec, in which $K$ is the temperature in kelvins. (The radius $R$ refers to the star's photosphere, the region where the star becomes effectively opaque to outside observation.) Stellar angular diameters can be measured through interferometry—that is, the combining of several telescopes together to form a larger instrument that can resolve sizes smaller than those that an individual telescope can resolve. Alternatively, the intensity of the starlight can be monitored during occultation by the Moon, which produces diffraction fringes whose pattern depends on the angular diameter of the star. Stellar angular diameters of several milliarcseconds can be measured.

Many stars occur in binary systems, in which the two partners orbit their mutual centre of mass. Such a system provides the best measurement of stellar masses. The period ($P$) of a binary system is related to the masses of the two stars ($m_1$ and $m_2$) and the orbital semimajor axis (mean radius; $a$) via Kepler's third law: $P^2 = 4\pi^2a^3/G(m_1 + m_2)$. ($G$ is the universal gravitational constant.) From diameters and masses, average values of the stellar density can be calculated and thence the central pressure. With the assumption of an equation of state, the central temperature can then be calculated. For example, in the Sun the central density is 158 grams per cubic cm; the pressure is calculated to be more than one billion times the pressure of Earth's atmosphere at sea level and the temperature around 15 million K (27 million °F). At this temperature, all atoms are ionized, and so the solar interior consists of a plasma, an ionized gas with hydrogen nuclei (i.e., protons), helium nuclei, and electrons as major constituents. A small fraction of the hydrogen nuclei possess sufficiently high speeds that, on colliding, their electrostatic repulsion is overcome, resulting in the formation, by means of a set of fusion reactions, of helium nuclei and a release of energy (*seeproton-proton* cycle). Some of this energy is carried away by neutrinos, but most of it is carried by photons to the surface of the Sun to maintain its luminosity.

Other stars, both more and less massive than the Sun, have broadly similar structures, but the size, central pressure and temperature, and fusion rate are functions of the star's mass and composition. The stars and their internal fusion (and resulting luminosity) are held stable against collapse through a delicate balance between the inward pressure produced by gravitational attraction and the outward pressure supplied by the photons produced in the fusion reactions.

Stars that are in this condition of hydrostatic equilibrium are termed main-sequence stars, and they occupy a well-defined band on the Hertzsprung-Russell (H-R) diagram, in which luminosity is plotted against colour index or temperature. Spectral classification, based initially on the colour index, includes the major spectral types O, B, A, F, G, K and M, each subdivided into 10 parts. Temperature is deduced from broadband spectral measurements in several standard wavelength intervals. Measurement of apparent magnitudes in two spectral regions, the $B$ and $V$ bands (centred on 4350 and 5550 angstroms, respectively), permits calculation of the colour index, CI = $m_B - m_V$, from which the temperature can be calculated.

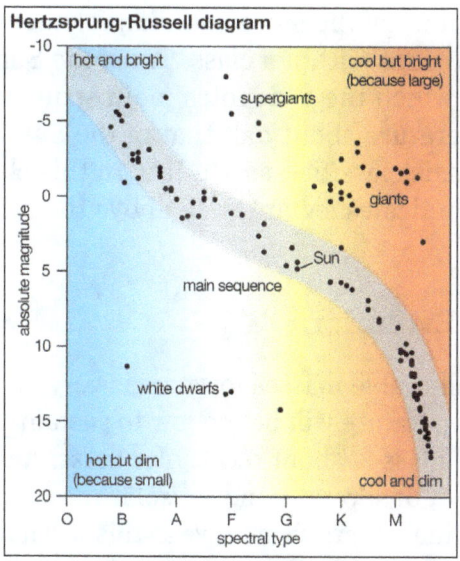

Hertzsprung-Russell diagram.

Hertzsprung-Russell diagram. Spectral type (a measure of a star's temperature), following the order introduced by American astronomer Annie Jump Cannon, is plotted on the horizontal axis, and absolute magnitude (the intrinsic brightness of a star) is plotted on the vertical axis.

For a given temperature, there are stars that are much more luminous than main-sequence stars. Given the dependence of luminosity on the square of the radius and the fourth power of the temperature ($R^2T^4$ of the luminosity expression above), greater luminosity implies larger radius, and such stars are termed giant stars or supergiant stars. Conversely, stars with luminosities much less than those of main-sequence stars of the same temperature must be smaller and are termed white dwarf stars. Surface temperatures of white dwarfs typically range from 10,000 to 12,000 K (18,000 to 21,000 °F), and they appear visually as white or blue-white.

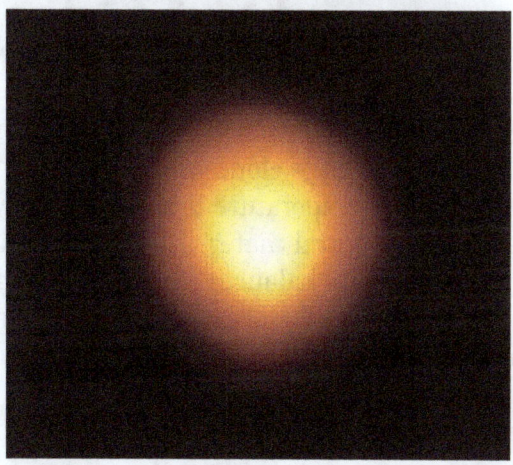

Betelgeuse imaged in ultraviolet light by the Hubble Space Telescope.

The strength of spectral lines of the more abundant elements in a star's atmosphere allows additional subdivisions within a class. Thus, the Sun, a main-sequence star, is classified as G2 V, in which the V denotes main sequence. Betelgeuse, a red giant with a surface temperature about half that of the Sun but with a luminosity of about 10,000 solar units, is classified as M2 Iab. In this classification, the spectral type is M2, and the Iab indicates a giant, well above the main sequence on the H-R diagram.

## Star formation and Evolution

The range of physically allowable masses for stars is very narrow. If the star's mass is too small, the central temperature will be too low to sustain fusion reactions. The theoretical minimum stellar mass is about 0.08 solar mass. An upper theoretical bound called the Eddington limit, of several hundred solar masses, has been suggested, but this value is not firmly defined. Stars as massive as this will have luminosities about one million times greater than that of the Sun.

Horsehead Nebula.

A general model of star formation and evolution has been developed, and the major features seem to be established. A large cloud of gas and dust can contract under its own gravitational attraction if its temperature is sufficiently low. As gravitational energy is released, the contracting central material heats up until a point is reached at which the outward radiation pressure balances the inward gravitational pressure, and contraction ceases. Fusion reactions take over as the star's primary source of energy, and the star is then on the main sequence. The time to pass through these formative stages and onto the main sequence is less than 100 million years for a star with as much mass as the Sun. It takes longer for less massive stars and a much shorter time for those much more massive.

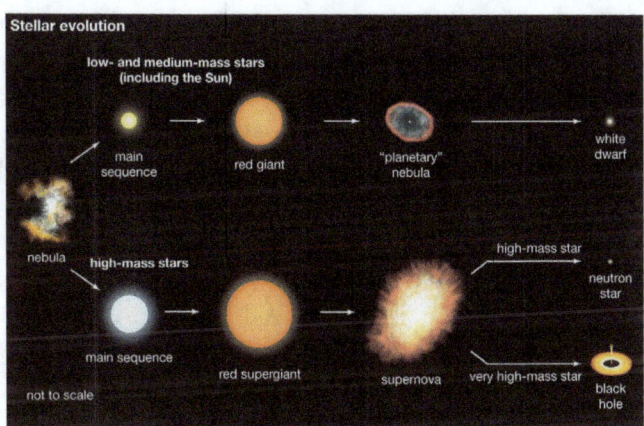

Stellar evolution.

Once a star has reached its main-sequence stage, it evolves relatively slowly, fusing hydrogen nuclei in its core to form helium nuclei. Continued fusion not only releases the energy that is radiated but also results in nucleosynthesis, the production of heavier nuclei.

Stellar evolution has of necessity been followed through computer modeling, because the timescales for most stages are generally too extended for measurable changes to be observed, even over a period of many years. One exception is the supernova, the violently explosive finale of certain stars. Different types of supernovas can be distinguished by their spectral lines and by changes in luminosity during and after the outburst. In Type Ia, a white dwarf star attracts matter from a nearby companion; when the white dwarf's mass exceeds about 1.4 solar masses, the star implodes and is completely destroyed. Type II supernovas are not as luminous as Type Ia and are the final evolutionary stage of stars more massive than about eight solar masses. Type Ib and Ic supernovas are like Type II in that they are from the collapse of a massive star, but they do not retain their hydrogen envelope.

The nature of the final products of stellar evolution depends on stellar mass. Some stars pass through an unstable stage in which their dimensions, temperature, and luminosity change cyclically over periods of hours or days. These so-called Cepheid variables serve as standard candles for distance measurements. Some stars blow off their outer layers to produce planetary nebulas. The expanding material can be seen glowing in a thin shell as it disperses into the interstellar medium while the remnant core, initially with a surface temperature as high as 100,000 K (180,000 °F), cools to become a white dwarf. The maximum stellar mass that can exist as a white dwarf is about 1.4 solar masses and is known as the Chandrasekhar limit. More-massive stars may end up as either neutron stars or black holes.

The average density of a white dwarf is calculated to exceed one million grams per cubic cm. Further compression is limited by a quantum condition called degeneracy, in which only certain energies are allowed for the electrons in the star's interior. Under sufficiently great pressure, the electrons are forced to combine with protons to form neutrons. The resulting neutron star will have a density in the range of 1014–1015 grams per cubic cm, comparable to the density within atomic nuclei. The behaviour of large masses having nuclear densities is not yet sufficiently understood to be able to set a limit on the maximum size of a neutron star, but it is thought to be less than three solar masses.

Nebula NGC 6751 formed several thousands of years ago as a dying star in the constellation Aquila threw off its outer layers of gas, which then fluoresced into a shell. The image, captured via NASA's Hubble Space Telescope, combines views taken through three different color filters in order to show the different temperatures of the nebular gases, with blue representing the hottest and orange and red the coolest. Our own sun will undergo a similar process in about 6 billion years, as it nears the end of its life.

Still more-massive remnants of stellar evolution would have smaller dimensions and would be even denser that neutron stars. Such remnants are conceived to be black holes, objects so compact that no radiation can escape from within a characteristic distance called the Schwarzschild radius. This critical dimension is defined by $R_s = 2GM/c^2$. (Rs is the Schwarzschild radius, G is the gravitational constant, M is the object's mass, and c is the speed of light.) For an object of three solar masses, the Schwarzschild radius would be about three kilometres. Radiation emitted from beyond the Schwarzschild radius can still escape and be detected.

Although no light can be detected coming from within a black hole, the presence of a black hole may be manifested through the effects of its gravitational field, as, for example, in a binary star system. If a black hole is paired with a normal visible star, it may pull matter from its companion toward itself. This matter is accelerated as it approaches the black hole and becomes so intensely heated that it radiates large amounts of X-rays from the periphery of the black hole before reaching the Schwarzschild radius. Some candidates for stellar black holes have been found—e.g., the X-ray source Cygnus X-1. Each of them has an estimated mass clearly exceeding that allowable for a neutron star, a factor crucial in the identification of possible black holes. Supermassive black holes that do not originate as individual stars exist at the centre of active galaxies. One such black hole, that at the center of the galaxy M87, has a mass 6.5 billion times that of the Sun and has been directly observed.

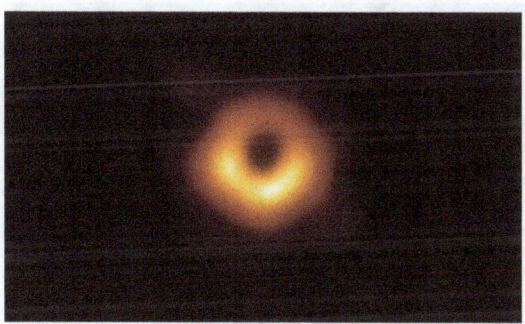

Black hole in M87.

Black hole at the centre of the massive galaxy M87, about 55 million light-years from Earth, as imaged by the Event Horizon Telescope (EHT). The black hole is 6.5 billion times more massive than the Sun. This image was the first direct visual evidence of a supermassive black hole and its shadow. The ring is brighter on one side because the black hole is rotating, and thus material on the side of the black hole turning toward Earth has its emission boosted by the Doppler effect. The shadow of the black hole is about five and a half times larger than the event horizon, the boundary marking the black hole's limits, where the escape velocity is equal to the speed of light.

Artist's conception of the X-ray source Cygnus X-1. A black hole pulls in matter from a stellar companion.

Whereas the existence of stellar black holes has been strongly indicated, the existence of neutron stars was confirmed in 1968 when they were identified with the then newly discovered pulsars, objects characterized by the emission of radiation at short and extremely regular intervals, generally between 1 and 1,000 pulses per second and stable to better than a part per billion. Pulsars are considered to be rotating neutron stars, remnants of some supernovas.

## Study of the Milky way Galaxy

Stars are not distributed randomly throughout space. Many stars are in systems consisting of two or three members separated by less than 1,000 AU. On a larger scale, star clusters may contain many thousands of stars. Galaxies are much larger systems of stars and usually include clouds of gas and dust.

Milky Way Galaxy as seen from Earth.

The solar system is located within the Milky Way Galaxy, close to its equatorial plane and about 8 kiloparsecs from the galactic centre. The galactic diameter is about 30 kiloparsecs, as indicated by luminous matter. There is evidence, however, for nonluminous matter—so-called dark matter—extending out nearly twice this distance. The entire system is rotating such that, at the position of the Sun, the orbital speed is about 220 km per second (almost 500,000 miles per hour) and a complete circuit takes roughly 240 million years. Application of Kepler's third law leads to an estimate for the galactic mass of about 100 billion solar masses. The rotational velocity can be measured from the Doppler shifts observed in the 21-cm emission line of neutral hydrogen and the lines of millimetre wavelengths from various molecules, especially carbon monoxide. At great distances from the galactic centre, the rotational velocity does not drop off as expected but rather increases slightly. This behaviour appears to require a much larger galactic mass than can be accounted for by the known (luminous) matter. Additional evidence for the presence of dark matter comes from a variety of other observations. The nature and extent of the dark matter (or missing mass) constitutes one of today's major astronomical puzzles.

There are about 100 billion stars in the Milky Way Galaxy. Star concentrations within the galaxy fall into three types: open clusters, globular clusters, and associations. Open clusters lie primarily in the disk of the galaxy; most contain between 50 and 1,000 stars within a region no more than 10 parsecs in diameter. Stellar associations tend to have somewhat fewer stars; moreover, the constituent stars are not as closely grouped as those in the clusters and are for the most part hotter. Globular clusters, which are widely scattered around the galaxy, may extend up to about 100 parsecs in diameter and may have as many as a million stars. The importance to astronomers of globular clusters lies in their use as indicators of the age of the galaxy. Because massive stars evolve more rapidly than do smaller stars, the age of a cluster can be estimated from its H-R diagram. In a young cluster the main sequence will be well populated, but in an old cluster the heavier stars will have evolved away from the main sequence. The extent of the depopulation of the main sequence provides an index of age. In this way, the oldest globular clusters have been found to be about 12.5 billion years old, which should therefore be the minimum age for the galaxy.

Globular cluster M80 (also known as NGC 6093) in an optical image taken by the Hubble Space Telescope. M80 is located 28,000 light-years from Earth and contains hundreds of thousands of stars.

## Investigations of Interstellar Matter

The interstellar medium, composed primarily of gas and dust, occupies the regions between the stars. On average, it contains less than one atom in each cubic centimetre, with about 1 percent of its mass in the form of minute dust grains. The gas, mostly hydrogen, has been mapped by means of its 21-cm emission line. The gas also contains numerous molecules. Some of these have been detected by the visible-wavelength absorption lines that they impose on the spectra of more-distant stars, while others have been identified by their own emission lines at millimetre wavelengths. Many of the interstellar molecules are found in giant molecular clouds, wherein complex organic molecules have been discovered.

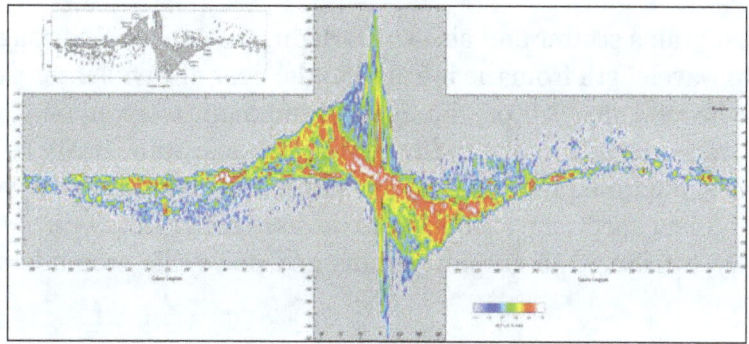

Longitude-velocity map of the Milky Way Galaxy as shown by spectral line emission of carbon monoxide in molecular clouds. The vertical axis represents velocity and the horizontal axis longitude. The gentle curves in the left and right portions of the map trace the spiral arms of the Milky Way Galaxy. The vertical structure in the middle of the map is the centre of the Galaxy. The emission stretching from the upper left to the lower right in the middle portion of the map is the "molecular ring," a ring of gas and dust in orbit between 4 and 8 kiloparsecs from the centre of the Galaxy.

In the vicinity of a very hot O- or B-type star, the intensity of ultraviolet radiation is sufficiently high to ionize the surrounding hydrogen out to a distance as great as 100 parsecs to produce an H II region, known as a Strömgren sphere. Such regions are strong and characteristic emitters of radiation at radio wavelengths, and their dimensions are well calibrated in terms of the luminosity of the central star. Using radio interferometers, astronomers are able to measure the angular diameters of H II regions even in some external galaxies and can thereby deduce the great distances to those remote systems. This method can be used for distances up to about 30 megaparsecs.

H II region: NGC 604, H II region inside the Triangulum Galaxy (M33),
photographed by Hubble Space Telescope. NGC 604 is one of the largest
H II regions; it contains more than 200 stars and is about 1,500 light-years in diameter.

Interstellar dust grains scatter and absorb starlight, the effect being roughly inversely proportional to wavelength from the infrared to the near ultraviolet. As a result, stellar spectra tend to be reddened. Absorption typically amounts to about one magnitude per kiloparsec but varies considerably in different directions. Some dusty regions contain silicate materials, identified by a broad absorption feature around a wavelength of 10 μm. Other prominent spectral features in the infrared range have been sometimes, but not conclusively, attributed to graphite grains and polycyclic aromatic hydrocarbons (PAHs).

Starlight often shows a small degree of polarization (a few percent), with the effect increasing with stellar distance. This is attributed to the scattering of the starlight from dust grains that have been partially aligned in a weak interstellar magnetic field. The strength of this field is estimated to be a few microgauss, very close to the strength inferred from observations of nonthermal cosmic radio noise. This radio background has been identified as synchrotron radiation, emitted by cosmic-ray electrons traveling at nearly the speed of light and moving along curved paths in the interstellar magnetic field. The spectrum of the cosmic radio noise is close to what is calculated on the basis of measurements of the cosmic rays near Earth.

Cosmic rays constitute another component of the interstellar medium. Cosmic rays that are detected in the vicinity of Earth comprise high-speed nuclei and electrons. Individual particle energies, expressed in electron volts (eV; $1 \text{ eV} = 1.6 \times 10^{-12}$ erg), range with decreasing numbers from about 106 eV to more than 1020 eV. Among the nuclei, hydrogen nuclei are the most plentiful at 86 percent, helium nuclei next at 13 percent, and all other nuclei together at about 1 percent. Electrons are about 2 percent as abundant as the nuclear component. (The relative numbers of different nuclei vary somewhat with kinetic energy, while the electron proportion is strongly energy-dependent.)

A minority of cosmic rays detected in Earth's vicinity are produced in the Sun, especially at times of increased solar activity (as indicated by sunspots and solar flares). The origin of galactic cosmic rays has not yet been conclusively identified, but they are thought to be produced in stellar processes such as supernova explosions, perhaps with additional acceleration occurring in the interstellar regions.

## Observations of the Galactic Centre

Cosmic radio-wave source Sagittarius A* in an image from the Chandra X-ray Observatory. Sagittarius A* is an extremely bright source within the larger Sagittarius A complex and contains the black hole at the Milky Way Galaxy's centre.

The central region of the Milky Way Galaxy is so heavily obscured by dust that direct observation has become possible only with the development of astronomy at nonvisual wavelengths—namely, radio, infrared, and, more recently, X-ray and gamma-ray wavelengths. Together, these observations have revealed a nuclear region of intense activity, with a large number of separate sources of emission and a great deal of dust. Detection of gamma-ray emission at a line energy of 511,000 eV, which corresponds to the annihilation of electrons and positrons (the antimatter counterpart of electrons), along with radio mapping of a region no more than 20 AU across, points to a very compact and energetic source, designated Sagittarius A*, at the centre of the galaxy. Sagittarius A* is a supermassive black hole with a mass equivalent to 4,310,000 Suns.

## Study of other Galaxies and Related Phenomena

Galaxies are normally classified into three principal types according to their appearance: spiral, elliptical, and irregular. Galactic diameters are typically in the tens of kiloparsecs and the distances between galaxies typically in megaparsecs.

Four irregular galaxies, as observed by the Hubble Space Telescope.

Spiral galaxies—of which the Milky Way system is a characteristic example—tend to be flattened, roughly circular systems with their constituent stars strongly concentrated along spiral arms. These arms are thought to be produced by traveling density waves, which compress and expand the galactic material. Between the spiral arms exists a diffuse interstellar medium of gas and dust, mostly at very low temperatures (below 100 K [−280 °F, −170 °C]). Spiral galaxies are typically a few kiloparsecs in thickness; they have a central bulge and taper gradually toward the outer edges.

Whirlpool Galaxy (M51); NGC 5195:The Whirlpool Galaxy (left), also known as M51, an Sc galaxy accompanied by a small irregular companion galaxy, NGC 5195 (right).

Ellipticals show none of the spiral features but are more densely packed stellar systems. They range in shape from nearly spherical to very flattened and contain little interstellar matter. Irregular galaxies number only a few percent of all stellar systems and exhibit none of the regular features associated with spirals or ellipticals.

The giant elliptical galaxy M87, also known as Virgo A, in an optical image taken by the Canada-France-Hawaii Telescope on Mauna Kea, Hawaii. M87 appears near the centre of the Virgo Cluster of galaxies.

Properties vary considerably among the different types of galaxies. Spirals typically have masses in the range of a billion to a trillion solar masses, with ellipticals having values from 10 times smaller to 10 times larger and the irregulars generally 10–100 times smaller. Visual galactic luminosities show similar spreads among the three types, but the irregulars tend to be less luminous. In contrast, at radio wavelengths the maximum luminosity for spirals is usually 100,000 times less than for ellipticals or irregulars.

3C 273, the brightest quasar, photographed by the Hubble Space Telescope's Advanced Camera for Surveys. The black region at the centre of the image is blocking light from the central object, revealing the host galaxy of 3C 273.

Quasars are objects whose spectra display very large redshifts, thus implying (in accordance with the Hubble law) that they lie at the greatest distances. They were discovered in 1963 but remained enigmatic for many years. They appear as star-like (i.e., very compact) sources of radio waves—hence their initial designation as quasi-stellar radio sources, a term later shortened to quasars. They are now considered to be the exceedingly luminous cores of distant galaxies. These energetic cores, which emit copious quantities of X-rays and gamma rays, are termed active galactic nuclei (AGN) and include the object Cygnus A and the nuclei of a class of galaxies called Seyfert galaxies. They are powered by the infall of matter into supermassive black holes.

The Milky Way Galaxy is one of the Local Group of galaxies, which contains about four dozen members and extends over a volume about two megaparsecs in diameter. Two of the closest members are the Magellanic Clouds, irregular galaxies about 50 kiloparsecs away. At about 740 kiloparsecs, the Andromeda Galaxy is one of the most distant in the Local Group. Some members of the group are moving toward the Milky Way system while others are traveling away from it. At greater distances, all galaxies are moving away from the Milky Way Galaxy. Their speeds (as determined from the redshifted wavelengths in their spectra) are generally proportional to their distances. The Hubble law relates these two quantities. In the absence of any other method, the Hubble law continues to be used for distance determinations to the farthest objects—that is, galaxies and quasars for which redshifts can be measured.

The Andromeda Galaxy, also known as the Andromeda Nebula or M31. It is the closest spiral galaxy to Earth, at a distance of 2.48 million light-years.

## Cosmology

Cosmology is the scientific study of the universe as a unified whole, from its earliest moments through its evolution to its ultimate fate. The currently accepted cosmological model is the big bang. In this picture, the expansion of the universe started in an intense explosion 13.8 billion years ago. In this primordial fireball, the temperature exceeded one trillion K, and most of the energy was in the form of radiation. As the expansion proceeded (accompanied by cooling), the role of the radiation diminished, and other physical processes dominated in turn. Thus, after about three minutes, the temperature had dropped to the one-billion-K range, making it possible for nuclear reactions of protons to take place and produce nuclei of deuterium and helium. (At the higher temperatures that prevailed earlier, these nuclei would have been promptly disrupted by high-energy photons.) With further expansion, the time between nuclear collisions had increased and the proportion of deuterium and helium nuclei had stabilized. After a few hundred thousand years, the temperature must have dropped sufficiently for electrons to remain attached to nuclei to constitute atoms. Galaxies are thought to have begun forming after a few million years, but this stage is very poorly understood. Star formation probably started much later, after at least a billion years, and the process continues today.

Observational support for this general model comes from several independent directions. The expansion has been documented by the redshifts observed in the spectra of galaxies. Furthermore, the radiation left over from the original fireball would have cooled with the expansion. Confirmation of this relic energy came in 1965 with one of the most striking cosmic discoveries of the 20th century—the observation, at short radio wavelengths, of a widespread cosmic radiation corresponding to a temperature of almost 3 K (about −270 °C [−454 °F]). The shape of the observed spectrum is an excellent fit with the theoretical Planck blackbody spectrum. (The present best value for this temperature is 2.735 K, but it is still called three-degree radiation or the cosmic microwave background.) The spectrum of this cosmic radio noise peaks at approximately a one-millimetre wavelength, which is in the far infrared, a difficult region to observe from Earth; however, the spectrum has been well mapped by the Cosmic Background Explorer (COBE), Wilkinson Microwave Anisotropy Probe, and Planck satellites. Additional support for the big bang theory comes from the observed cosmic abundances of deuterium and helium. Normal stellar nucleosynthesis cannot produce their measured quantities, which fit well with calculations of production during the early stages of the big bang.

Early surveys of the cosmic background radiation indicated that it is extremely uniform in all directions (isotropic). Calculations have shown that it is difficult to achieve this degree of isotropy unless there was a very early and rapid inflationary period before the expansion settled into its present mode. Nevertheless, the isotropy posed problems for models of galaxy formation. Galaxies originate from turbulent conditions that produce local fluctuations of density, toward which more matter would then be gravitationally attracted. Such density variations were difficult to reconcile with the isotropy required by observations of the 3 K radiation. This problem was solved when the COBE satellite was able to detect the minute fluctuations in the cosmic background from which the galaxies formed.

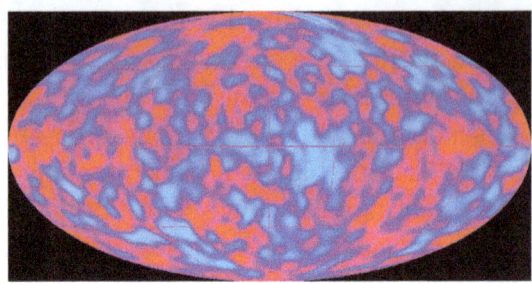

Image of the cosmic microwave background, taken by the Differential Microwave Radiometer on board the U.S. satellite Cosmic Background Explorer. The red features in the image show places where the universe was slightly denser, thus stimulating gravitational separation and, ultimately, the formation of galaxies.

The very earliest stages of the big bang are less well understood. The conditions of temperature and pressure that prevailed prior to the first microsecond require the introduction of theoretical ideas of subatomic particle physics. Subatomic particles are usually studied in laboratories with giant accelerators, but the region of particle energies of potential significance to the question at hand lies beyond the range of accelerators currently available. Fortunately, some important conclusions can be drawn from the observed cosmic helium abundance, which is dependent on conditions in the early big bang. The observed helium abundance sets a limit on the number of families of certain types of subatomic particles that can exist.

The age of the universe can be calculated in several ways. Assuming the validity of the big bang model, one attempts to answer the question: How long has the universe been expanding in order to have reached its present size? The numbers relevant to calculating an answer are Hubble's constant (i.e., the current expansion rate), the density of matter in the universe, and the cosmological constant, which allows for change in the expansion rate. In 2003 a calculation based on a fresh determination of Hubble's constant yielded an age of 13.7 billion ± 200 million years, although the precise value depends on certain assumed details of the model used. Independent estimates of stellar ages have yielded values less than this, as would be expected, but other estimates, based on supernova distance measurements, have arrived at values of about 15 billion years, still consistent, within the errors. In the big bang model the age is proportional to the reciprocal of Hubble's constant, hence the importance of determining H as reliably as possible. For example, a value for H of 100 km/sec/Mpc would lead to an age less than that of many stars, a physically unacceptable result.

A small minority of astronomers have developed alternative cosmological theories that are seriously pursued. The overwhelming professional opinion, however, continues to support the big bang model.

Finally, there is the question of the future behaviour of the universe: Is it open? That is to say, will the expansion continue indefinitely? Or is it closed, such that the expansion will slow down and eventually reverse, resulting in contraction? (The final collapse of such a contracting universe is sometimes termed the "big crunch.") The density of the universe seems to be at the critical density; that is, the universe is neither open nor closed but "flat." So-called dark energy, a kind of repulsive force that is now believed to be a major component of the universe, appears to be the decisive factor in predictions of the long-term fate of the cosmos. If this energy is a cosmological constant (as proposed in 1917 by Albert Einstein to correct certain problems in his model of the universe), then the result would be a "big chill." In this scenario, the universe would continue to expand, but its density would decrease. While old stars would burn out, new stars would no longer form. The universe would become cold and dark. The dark (nonluminous) matter component of the universe, whose composition remains unknown, is not considered sufficient to close the universe and cause it to collapse; it now appears to contribute only a fourth of the density needed for closure.

An additional factor in deciding the fate of the universe might be the mass of neutrinos. For decades the neutrino had been postulated to have zero mass, although there was no compelling theoretical reason for this to be so. From the observation of neutrinos generated in the Sun and other celestial sources such as supernovas, in cosmic-ray interactions with Earth's atmosphere, and in particle accelerators, investigators have concluded that neutrinos have some mass, though only an extremely small fraction of the mass of an electron. Although there are vast numbers of neutrinos in the universe, the sum of such small neutrino masses appears insufficient to close the universe.

## The Techniques of Astronomy

Astronomical observations involve a sequence of stages, each of which may impose constraints on the type of information attainable. Radiant energy is collected with telescopes and brought to a focus on a detector, which is calibrated so that its sensitivity and spectral response are known. Accurate pointing and timing are required to permit the correlation of observations made with different instrument systems working in different wavelength intervals and located at places far apart. The radiation must be spectrally analyzed so that the processes responsible for radiation emission can be identified.

## Telescopic Observations

Before Galileo Galilei's use of telescopes for astronomy in 1609, all observations were made by naked eye, with corresponding limits on the faintness and degree of detail that could be seen. Since that time, telescopes have become central to astronomy. Having apertures much larger than the pupil of the human eye, telescopes permit the study of faint and distant objects. In addition, sufficient radiant energy can be collected in short time intervals to permit rapid fluctuations in intensity to be detected. Further, with more energy collected, a spectrum can be greatly dispersed and examined in much greater detail.

Aerial view of the Keck Observatory's twin domes, which are opened to reveal the telescopes. Keck II is on the left and Keck I on the right.

Optical telescopes are either refractors or reflectors that use lenses or mirrors, respectively, for their main light-collecting elements (objectives). Refractors are effectively limited to apertures of about 100 cm (approximately 40 inches) or less because of problems inherent in the use of large glass lenses. These distort under their own weight and can be supported only around the perimeter; an appreciable amount of light is lost due to absorption in the glass. Large-aperture refractors are very long and require large and expensive domes. The largest modern telescopes are all reflectors, the very largest composed of many segmented components and having overall diameters of about 10 metres (33 feet). Reflectors are not subject to the chromatic problems of refractors, can be better supported mechanically, and can be housed in smaller domes because they are more compact than the long-tube refractors.

The angular resolving power (or resolution) of a telescope is the smallest angle between close objects that can be seen clearly to be separate. Resolution is limited by the wave nature of light. For a telescope having an objective lens or mirror with diameter D and operating at wavelength $\lambda$, the angular resolution (in radians) can be approximately described by the ratio $\lambda/D$. Optical telescopes can have very high intrinsic resolving powers; in practice, however, these are not attained for telescopes located on Earth's surface, because atmospheric effects limit the practical resolution to about one arc second. Sophisticated computing programs can allow much-improved resolution, and the performance of telescopes on Earth can be improved through the use of adaptive optics, in which the surface of the mirror is adjusted rapidly to compensate for atmospheric turbulence that would otherwise distort the image. In addition, image data from several telescopes focused on the same object can be merged optically and through computer processing to produce images having angular resolutions much greater than that from any single component.

The atmosphere does not transmit radiation of all wavelengths equally well. This restricts astronomy on Earth's surface to the near ultraviolet, visible, and radio regions of the electromagnetic spectrum and to some relatively narrow "windows" in the nearer infrared. Longer infrared wavelengths are strongly absorbed by atmospheric water vapour and carbon dioxide. Atmospheric effects can be reduced by careful site selection and by carrying out observations at high altitudes. Most major optical observatories are located on high mountains, well away from cities and their reflected lights. Infrared telescopes have been located atop Mauna Kea in Hawaii, in the Atacama Desert in Chile, and in the Canary Islands, where atmospheric humidity is very low. Airborne telescopes designed mainly for infrared observations—such as on the Stratospheric Observatory for Infrared Astronomy (SOFIA), a jet aircraft fitted with astronomical instruments—operate at an altitude of about 12 km (40,000 feet) with flight durations limited to a few hours. Telescopes for infrared, X-ray, and gamma-ray observations have been carried to altitudes of more than 30 km (100,000 feet) by balloons. Higher altitudes can be attained during short-duration rocket flights for ultraviolet observations. Telescopes for all wavelengths from infrared to gamma rays have been carried by robotic spacecraft observatories such as the Hubble Space Telescope and the Wilkinson Microwave Anisotropy Probe, while cosmic rays have been studied from space by the Advanced Composition Explorer.

The James Clerk Maxwell Telescope located near the summit of Mauna Kea, Hawaii.

Angular resolution better than one milliarcsecond has been achieved at radio wavelengths by the use of several radio telescopes in an array. In such an arrangement, the effective aperture then becomes the greatest distance between component telescopes. For example, in the Very Large Array (VLA), operated near Socorro, New Mexico, by the National Radio Astronomy Observatory, 27 movable radio dishes are set out along tracks that extend for nearly 21 km. In another technique, called very long baseline interferometry (VLBI), simultaneous observations are made with radio telescopes thousands of kilometres apart; this technique requires very precise timing.

Very Large Array, radio telescope system located on the plains of San Agustin, near Socorro, New Mexico.

Earth is a moving platform for astronomical observations. It is important that the specification of precise celestial coordinates be made in ways that correct for telescope location, the position of Earth in its orbit around the Sun, and the epoch of observation, since Earth's axis of rotation moves slowly over the years. Time measurements are now based on atomic clocks rather than on Earth's rotation, and telescopes can be driven continuously to compensate for the planet's rotation, so as to permit tracking of a given astronomical object.

## Use of Radiation Detectors

Although the human eye remains an important astronomical tool, detectors capable of greater sensitivity and more rapid response are needed to observe at visible wavelengths and, especially, to extend observations beyond that region of the electromagnetic spectrum. Photography was an essential tool from the late 19th century until the 1980s, when it was supplanted by charge-coupled devices (CCDs). However, photography still provides a useful archival record. A photograph of a particular celestial object may include the images of many other objects that were not of interest when the picture was taken but that become the focus of study years later. When quasars were discovered in 1963, for example, photographic plates exposed before 1900 and held in the Harvard College Observatory were examined to trace possible changes in position or intensity of the radio object newly identified as quasar 3C 273. Also, major photographic surveys, such as those of the National Geographic Society and the Palomar Observatory, can provide a historical base for long-term studies.

Photographic film converted only a few percent of the incident photons into images, whereas CCDs have efficiencies of nearly 100 percent. CCDs can be used for a wide range of wavelengths, from the X-ray into the near-infrared. Gamma rays are detectable through their Compton scattering, electron-positron pair production, or Cerenkov radiation. For infrared wavelengths longer than a few microns, semiconductor detectors that operate at very low (cryogenic) temperatures are used. Reception of radio waves is based on the production of a small voltage in an antenna rather than on photon counting.

Spectroscopy involves measuring the intensity of the radiation as a function of wavelength or frequency. In some detectors, such as those for X-rays and gamma rays, the energy of each photon can be measured directly. For low-resolution spectroscopy, broadband filters suffice to select wavelength intervals. Greater resolution can be obtained with prisms, gratings, and interferometers.

## Multi-messenger Astronomy

Most of what is known about the universe comes from observations of electromagnetic radiation. However, there are other "cosmic messengers." Gravity waves are disturbances in space-time that can be detected by very large laser interferometers. Gravity waves and gamma-ray bursts have been observed from neutron-star mergers.

Neutrinos and cosmic rays are other particles that can, in principle, be observed; however, as yet, these latter messengers cannot be identified with specific sources. Using two or more of these methods is called multi-messenger astronomy.

The Laser Interferometer Gravitational-Wave Observatory (LIGO) near Hanford, Washington, U.S. There are two LIGO installations; the other is near Livingston, Louisiana, U.S.

## Solid Cosmic Samples

As a departure from the traditional astronomical approach of remote observing, certain more recent lines of research involve the analysis of actual samples under laboratory conditions. These include studies of meteorites, rock samples returned from the Moon, cometary and asteroid dust samples returned by space probes, and interplanetary dust particles collected by aircraft in the stratosphere or by spacecraft. In all such cases, a wide range of highly sensitive laboratory techniques can be adapted for the often microscopic samples. Chemical analysis can be supplemented with mass spectrometry, allowing isotopic composition to be determined. Radioactivity and the impacts of cosmic-ray particles can produce minute quantities of gas, which then remain trapped in crystals within the samples. Carefully controlled heating of the crystals (or of dust grains containing the crystals) under laboratory conditions releases this gas, which then is analyzed in a mass spectrometer. X-ray spectrometers, electron microscopes, and microprobes are employed to determine crystal structure and composition, from which temperature and pressure conditions at the time of formation can be inferred.

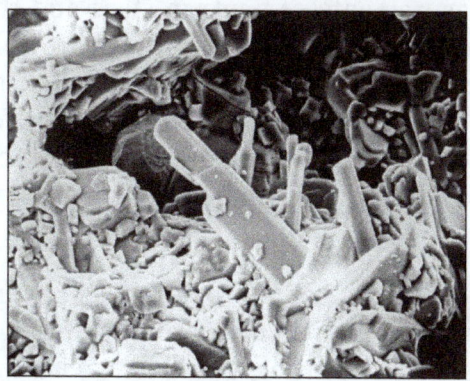

Moon rock; crystals.

A scanning-electron-microscope photograph of pyroxene and plagioclase crystals (the long and the short crystals, respectively) that grew in a cavity in a fragment of Moon rock gathered during the Apollo 14 mission.

## Theoretical Approaches

Theory is just as important as observation in astronomy. It is required for the interpretation of observational data; for the construction of models of celestial objects and physical processes, their properties, and their changes over time; and for guiding further observations. Theoretical astrophysics is based on laws of physics that have been validated with great precision through controlled experiments. Application of these laws to specific astrophysical problems, however, may yield equations too complex for direct solution. Two general approaches are then available. In the traditional method, a simplified description of the problem is formulated, incorporating only the major physical components, to provide equations that can be either solved directly or used to create a numerical model that can be evaluated. Successively more-complex models can then be investigated. Alternatively, a computer program can be devised that will explore the problem numerically in all its complexity. Computational science has taken its place as a major division alongside theory and experiment. The test of any theory is its ability to incorporate the known facts and to make predictions that can be compared with additional observations.

## Impact of Astronomy

No area of science is totally self-contained. Discoveries in one area find applications in others, often unpredictably. Various notable examples of this involve astronomical studies. Isaac Newton's laws of motion and gravity emerged from the analysis of planetary and lunar orbits. Observations during the 1919 solar eclipse provided dramatic confirmation of Albert Einstein's general theory of relativity, which gained further support with the discovery of the binary pulsar designated PSR 1913+16 and the observation of gravity waves from merging black holes and neutron stars. The behaviour of nuclear matter and of some elementary particles is now better understood as a result of measurements of neutron stars and the cosmological helium abundance, respectively. Study of the theory of synchrotron radiation was greatly stimulated by the detection of polarized visible radiation emitted by high-energy electrons in the supernova remnant known as the Crab Nebula. Dedicated particle accelerators are now being used to produce synchrotron radiation to probe the structure of solid materials and make detailed X-ray images of tiny samples, including biological structures.

Astronomical knowledge also has had a broad impact beyond science. The earliest calendars were based on astronomical observations of the cycles of repeated solar and lunar positions. Also, for centuries, familiarity with the positions and apparent motions of the stars through the seasons enabled sea voyagers to navigate with moderate accuracy. Perhaps the single greatest effect that astronomical studies have had on our modern society has been in molding its perceptions and opinions. Our conceptions of the cosmos and our place in it, our perceptions of space and time, and the development of the systematic pursuit of knowledge known as the scientific method have been profoundly influenced by astronomical observations. In addition, the power of science to provide the basis for accurate predictions of such phenomena as eclipses and the positions of the planets and later, so dramatically, of comets has shaped an attitude toward science that remains an important social force today.

## Astrophysics

Astrophysics is a branch of space science that applies the laws of physics and chemistry to explain the birth, life and death of stars, planets, galaxies, nebulae and other objects in the universe. It has two sibling sciences, astronomy and cosmology, and the lines between them blur.

In the most rigid sense:

- Astronomy measures positions, luminosities, motions and other characteristics.

- Astrophysics creates physical theories of small to medium-size structures in the universe.

- Cosmology does this for the largest structures, and the universe as a whole.

### Goals of Astrophysics

Astrophysicists seek to understand the universe and our place in it. At NASA, the goals of astrophysics are "to discover how the universe works, explore how it began and evolved, and search for life on planets around other stars.

NASA states that those goals produce three broad questions:

- How does the universe work?

- How did we get here?

- Are we alone?

### It Began with Newton

While astronomy is one of the oldest sciences, theoretical astrophysics began with Isaac Newton. Prior to Newton, astronomers described the motions of heavenly bodies using complex mathematical models without a physical basis. Newton showed that a single theory simultaneously explains the orbits of moons and planets in space and the trajectory of a cannonball on Earth. This added to the body of evidence for the (then) startling conclusion that the heavens and Earth are subject to the same physical laws.

Perhaps what most completely separated Newton's model from previous ones is that it is predictive as well as descriptive. Based on aberrations in the orbit of Uranus, astronomers predicted the position of a new planet, which was then observed and named Neptune. Being predictive as well as descriptive is the sign of a mature science, and astrophysics is in this category.

Because the only way we interact with distant objects is by observing the radiation they emit, much of astrophysics has to do with deducing theories that explain the mechanisms that produce this radiation, and provide ideas for how to extract the most information from it. The first ideas about the nature of stars emerged in the mid-19th century from the blossoming science of spectral analysis, which means observing the specific frequencies of light that particular substances absorb and emit when heated. Spectral analysis remains essential to the triumvirate of space sciences, both guiding and testing new theories.

Early spectroscopy provided the first evidence that stars contain substances also present on Earth. Spectroscopy revealed that some nebulae are purely gaseous, while some contain stars. This later helped cement the idea that some nebulae were not nebulae at all — they were other galaxies.

In the early 1920s, Cecilia Payne discovered, using spectroscopy, that stars are predominantly hydrogen (at least until their old age). The spectra of stars also allowed astrophysicists to determine the speed at which they move toward or away from Earth. Just like the sound a vehicle emits is different moving toward us or away from us, because of the Doppler shift, the spectra of stars will change in the same way. In the 1930s, by combining the Doppler shift and Einstein's theory of general relativity, Edwin Hubble provided solid evidence that the universe is expanding. This is also predicted by Einstein's theory, and together form the basis of the Big Bang Theory.

Also in the mid-19th century, the physicists Lord Kelvin (William Thomson) and Gustav Von Helmholtz speculated that gravitational collapse could power the sun, but eventually realized that energy produced this way would only last 100,000 years. Fifty years later, Einstein's famous $E=mc^2$ equation gave astrophysicists the first clue to what the true source of energy might be (although it turns out that gravitational collapse does play an important role). As nuclear physics, quantum mechanics and particle physics grew in the first half of the 20th century, it became possible to formulate theories for how nuclear fusion could power stars. These theories describe how stars form, live and die, and successfully explain the observed distribution of types of stars, their spectra, luminosities, ages and other features.

Astrophysics is the physics of stars and other distant bodies in the universe, but it also hits close to home. According to the Big Bang Theory, the first stars were almost entirely hydrogen. The nuclear fusion process that energizes them smashes together hydrogen atoms to form the heavier element helium. In 1957, the husband-and-wife astronomer team of Geoffrey and Margaret Burbidge, along with physicists William Alfred Fowler and Fred Hoyle, showed how, as stars age, they produce heavier and heavier elements, which they pass on to later generations of stars in ever-greater quantities. It is only in the final stages of the lives of more recent stars that the elements making up the Earth, such as iron (32.1 percent), oxygen (30.1 percent), silicon (15.1 percent), are produced. Another of these elements is carbon, which together with oxygen, make up the bulk of the mass of all living things, including us. Thus, astrophysics tells us that, while we are not all stars, we are all stardust.

# Branches of Astrophysics 2

- **Computational Astrophysics**
- **Astroparticle Physics**
- **Nuclear Astrophysics**
- **Atomic and Molecular Astrophysics**
- **Solar Physics**
- **Astrophysical Fluid Dynamics**
- **Stellar Dynamics**

Some of the various branches of astrophysics include solar physics, astroparticle physics, atomic and molecular astrophysics, computational astrophysics, nuclear astrophysics, astrophysical fluid dynamics and stellar dynamics. This chapter has been carefully written to provide an easy understanding of these branches of astrophysics.

## Computational Astrophysics

Computational astrophysics is the use of numerical methods to solve research problems in astrophysics on a computer.

Numerical methods are used whenever the mathematical model describing an astrophysical system is too complex to solve analytically (with pencil and paper). Today, it is difficult to find examples of research problems that do *not* use computation.

### Computational versus Analytic Methods

Solutions generated by numerical methods are generally only approximations to the exact solution of the underlying equations. However, much more complex systems of equations can be solved numerically than can be solved analytically. Thus, approximate solutions to the exact equations found by numerical methods often provide far more insight than exact solutions to approximate equations that can be solved analytically. For example, time-dependent numerical solutions of fluid flow in three dimensions can exhibit behavior which is not expected from one-dimensional analytic solutions to the steady-state (time independent) equations.

The increase in computing power in the last few decades has meant that an increasingly larger share of problems in astrophysics can be solved on a desktop computer. However, the most computationally intensive problems in astrophysics are still limited by the memory and floating-point speed of the largest high-performance computer (HPC) systems available (e.g. computers on the top500 list). The use of HPC is necessary to maximize the spatial, temporal, or frequency resolution of solutions, to include more physics, or when large parameter surveys are required to understand the statistics.

Examples:

Traditionally, the most important applications of computation in astrophysics have been in the areas of:

- Stellar structure and evolution.

  The internal structure and evolution of stars with different masses and chemical composition was largely mapped out in the 1960s using numerical methods to solve the equations of stellar structure. Today, the frontiers of research include calculating multidimensional stellar models of rapidly rotating stars, modeling the effects of hydrodynamical processes such as convection from first principles, and understanding how stars generate magnetic fields through dynamo processes.

- Radiation transfer and stellar atmospheres.

  Without oversimplifying assumptions, computational methods are required to calculate the propagation of light through the outer layers of a star, including its interaction with matter through absorption, emission, and scattering of photons. The calculation of cross sections for the interaction of light with matter for astrophysically relevant ions is itself a challenging computational problem. The construction of such stellar atmosphere models share many challenges with radiation transfer problems in other systems such as planets, accretion disks, and the interstellar medium. Modern calculations improve the frequency resolution, include a better treatment of opacities, and can treat non-hydrostatic atmospheres.

- Astrophysical fluid dynamics.

  The dynamics of most of the visible matter in the universe can be treated as a compressible fluid. Time-dependent and multidimensional solutions to the fluid equations, including the effects of gravitational, magnetic, and radiation fields, require numerical methods. A vast range of problems are addressed in this way, from convection and dynamo action in stellar and planetary interiors, to the formation of galaxies and the large scale structure of the universe.

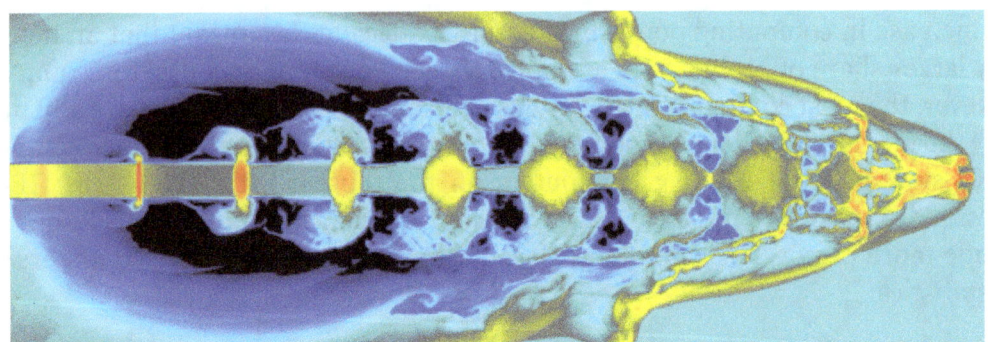

Density in a 2D simulations of the propagation of a pulsed, supersonic, protostellar jet.

- Planetary, stellar, and galactic dynamics.

  It is well known that Newton's Laws of Motion for some number of particles $N$ interacting through their mutual gravitational attraction do not in general have an analytic solution for $N > 2$. Thus, to compute the orbits of planets in the solar system, or of stars in the Galaxy, numerical methods are required. The most challenging problems today include accurate integration of the orbits of the planets over the age of the solar system, studying the dynamics of globular clusters including the effect of stellar evolution and the formation of binaries, studying galaxy mergers and interaction, and computing structure formation in the universe through the gravitational clustering of collisionless dark matter.

## Numerical Methods

The diverse set of mathematical models encountered in astrophysics means that a very wide range of numerical methods are necessary. These range from basic methods for linear algebra, nonlinear root finding, and ordinary differential equations (ODEs), to more complex methods for coupled partial differential equations (PDEs) in multidimensions, topics which span the entire content of reference books such as Numerical Recipes.

However, there are several numerical methods used in astrophysics that deserve special mention, either because they have such wide use in astrophysics, or because astrophysicists have made significant contributions to their development. These methods are considered in the following subsections.

## Stellar Structure Codes

The equations of stellar structure are a set of ODEs which define a classic two-point boundary value problem. Analytic solutions to an approximate system of equations exist in special cases (*polytropes*) but these are of limited application to real stars. Numerical solutions to the full system of equations were first computed using *shooting methods*, in which boundary conditions are guessed at the center and surface of the star, the equations are integrated outwards and inwards, and matching conditions are used at some interior point to select the full solution. Shooting methods are laborious and cumbersome; today modern stellar structure codes use relaxation schemes which find the solution to the finite-difference form of the stellar structure equations by finding the roots of coupled, non-linear equations at each mesh point. A good example of a public code that uses a relaxation scheme is the EZ-code, based on Eggleton's variable mesh method. The evolution of stars are then computed by computing stellar models at discrete time intervals, with the chemical composition of the star modified by nuclear reactions in the interior.

## Radiative Transfer Codes

Calculating the emergent intensity from an astrophysical system requires solving a multidimensional integral-differential equation, along with level population equations to account for the interaction of the radiation with matter. In general the solution is a function of two angles, frequency, and time. Even in static, plane-parallel atmospheres, the problem is two-dimensional (one angle and frequency). However, the most challenging aspect of the problem is that scattering couples the solutions at different angles and frequencies. As in the stellar structure problem, relaxation schemes are used to solve the finite difference form of the transfer equations, although specialized iteration techniques are necessary to accelerate convergence. Monte-Carlo methods, which adopt statistical techniques to approximate the solution by following the propagation of many photon packets, are becoming increasingly important. The problem of line-transfer in a moving atmosphere (stellar wind) is especially challenging, due to non-local coupling introduced by Doppler shifts in the spectrum.

## N-body Codes

There are two tasks in an N-body code: integrating the equations of motion (*pushing particles*), and computing the gravitational acceleration of each particle. The former requires methods for integrating ODEs. Modern codes are based on a combination of high-order difference approximations (e.g. Hermite integrators), and symplectic methods (which have the important property of generating solutions that obey Liouville's Theorem, i.e. that preserve the volume of the solution in phase space). Symplectic methods are especially important for long term time integration, because they control the accumulation of truncation error in the solution.

Calculation of the gravitational acceleration is challenging because the computational cost scales as $N(N-1)$, where $N$ is the number of particles. For small $N$, direct summation can be used. For moderate $N$(currently $N \sim 10^{5-6}$), special purpose hardware (e.g. GRAPE boards) can be used to accelerate the evaluation of $1/r^2$ necessary to compute the acceleration by direct summation. Finally, for large $N$ (currently $N \geq 10^{9-10}$), tree-methods are used to approximate the force from distant particles.

## Codes for Astrophysical Fluid Dynamics

Solving the equations of compressible gas dynamics is a classic problem in numerical analysis which has application to many fields besides astrophysics. Thus, a large number of methods have been developed, with many important contributions being made by astrophysicists. For solving the equations of compressible fluid dynamics, the most popular methods include:

- Finite-difference techniques (which require hyper-viscosity to smooth discontinuities),

- Finite-volume methods (which often use a Riemann solver to compute upwind fluxes),

- Operator split methods (which combine elements of both finite-differencing and finite-volume methods for different terms in the equations),

- Central methods (which often use simple expressions for the fluxes, combined with high-order spatial interpolation),

- Particle methods such as smooth particle hydrodynamics (SPH, which integrates the motion of discrete particles to follow the flow).

A good technical review of many of these methods is given by LeVeque (2002). SPH is an example of a method developed largely to solve astrophysics problems, although many of the developments in other methods (e.g. the extension of finite-difference and finite-volume methods to magnetohydrodynamics (MHD) to include the effects of magnetic fields on the dynamics of the fluid) have also been motivated by astrophysics.

## Relation to other Fields

Computational astrophysics is necessarily inter-disciplinary, involving aspects of not only astrophysics, but also numerical analysis and computer science.

## Numerical Analysis

Numerical Analysis is a rigorous branch of mathematics concerned with the approximation of functions and integrals, and the approximation of solutions to algebraic, differential, and integral equations. It provides tools to analyze errors that arise from the approximations themselves (truncation error), and from the use of finite-precision arithmetic on a computer (round-off error). Convergence, consistency, and stability of numerical algorithms are all essential for their use in practical applications. Thus, the development of new numerical algorithms to solve problems in astrophysics is deeply rooted in the tools of numerical analysis.

### Computer Science

Computational science and computer science differ. The former is the use of numerical methods to solve scientific problems. The latter is the study of computers and computation. Thus, computer science tends to be more orientated towards the theory of computers and computation, while scientific computation is concerned with the practical aspects of solving science problems. Still, there is a large degree of overlap between the fields, with many computer scientists engaged in developing tools for scientific computation, and many scientists working on software problems of interest to computer scientists. Examples of the overlap include the development of standards for parallel processing such as the Message Passing Interface (MPI) and OpenMP, and development of parallel I/O filesystems such as Lustre.

## Astroparticle Physics

Astroparticle physics, also called particle astrophysics, is a branch of particle physics that studies elementary particles of astronomical origin and their relation to astrophysics and cosmology. It is a relatively new field of research emerging at the intersection of particle physics, astronomy, astrophysics, detector physics, relativity, solid state physics, and cosmology. Partly motivated by the discovery of neutrino oscillation, the field has undergone rapid development, both theoretically and experimentally, since the early 2000s.

The field of astroparticle physics is evolved out of optical astronomy. With the growth of detector technology came the more mature astrophysics, which involved multiple physics subtopics, such as mechanics, electrodynamics, thermodynamics, plasma physics, nuclear physics, relativity, and particle physics. Particle physicists found astrophysics necessary due to difficulty in producing particles with comparable energy to those found in space. For example, the cosmic ray spectrum contains particles with energies as high as $10^{20}$ eV, where a proton-proton collision at the Large Hadron Collider occurs at an energy of ~$10^{12}$ eV.

The field can be said to have begun in 1910, when a German physicist named Theodor Wulf measured the ionization in the air, an indicator of gamma radiation, at the bottom and top of the Eiffel Tower. He found that there was far more ionization at the top than what was expected if only terrestrial sources were attributed for this radiation.

The Austrian physicist Victor Francis Hess hypothesized that some of the ionization was caused by radiation from the sky. In order to defend this hypothesis, Hess designed instruments capable of operating at high altitudes and performed observations on ionization up to an altitude of 5.3 km. From 1911 to 1913, Hess made ten flights to meticulously measure ionization levels. Through prior calculations, he did not expect there to be any ionization above an altitude of 500 m if terrestrial sources were the sole cause of radiation. His measurements however, revealed that although the ionization levels initially decreased with altitude, they began to sharply rise at some point. At the peaks of his flights, he found that the ionization levels were much greater than at the surface. Hess was then able to conclude that "a radiation of very high penetrating power enters our atmosphere from above." Furthermore, one of Hess's flights was during a near-total eclipse of the Sun. Since he did not observe a dip in ionization levels, Hess reasoned that the source had to be further away in space. For this discovery, Hess was one of the people awarded the Nobel Prize in Physics in 1936. In 1925, Robert Millikan confirmed Hess's findings and subsequently coined the term 'cosmic rays'.

Many physicists knowledgeable about the origins of the field of astroparticle physics prefer to attribute this 'discovery' of cosmic rays by Hess as the starting point for the field.

While it may be difficult to decide on a standard 'textbook' description of the field of astroparticle physics, the field can be characterized by the topics of research that are actively being pursued. The journal *Astroparticle Physics* accepts papers that are focused on new developments in the following areas:

- High-energy cosmic-ray physics and astrophysics;

- Particle cosmology;

- Particle astrophysics;

- Related astrophysics: Supernova, Active Galactic Nuclei, Cosmic Abundances, Dark Matter etc.;

- High-energy, VHE and UHE gamma-ray astronomy;

- High- and low-energy neutrino astronomy;

- Instrumentation and detector developments related to the above-mentioned fields.

One main task for the future of the field is simply to thoroughly define itself beyond working definitions and clearly differentiate itself from astrophysics and other related topics.

Current unsolved problems for the field of astroparticle physics include characterization of dark matter and dark energy. Observations of the orbital velocities of stars in the Milky Way and other galaxies starting with Walter Baade and Fritz Zwicky in the 1930s, along with observed velocities of galaxies in galactic clusters, found motion far exceeding the energy density of the visible matter needed to account for their dynamics. Since the early nineties some candidates have been found to partially explain some of the missing dark matter, but they are nowhere near sufficient to offer a full explanation. The finding of an accelerating universe suggests that a large part of the missing dark matter is stored as dark energy in a dynamical vacuum.

Another question for astroparticle physicists is why is there so much more matter than antimatter in the universe today. Baryogenesis is the term for the hypothetical processes that produced the unequal numbers of baryons and anitbaryons in the early universe, which is why the universe is made of matter today, and not antimatter.

## Experimental Facilities

The rapid development of this field has led to the design of new types of infrastructure. In underground laboratories or with specially designed telescopes, antennas and satellite experiments, astroparticle physicists employ new detection methods to observe a wide range of cosmic particles including neutrinos, gamma rays and cosmic rays at the highest energies. They are also searching for dark matter and gravitational waves. Experimental particle physicists are limited by the technology of their terrestrial accelerators, which are only able to produce a small fraction of the energies found in nature.

Facilities, experiments and laboratories involved in astroparticle physics include:

- IceCube (Antarctica). The longest particle detector in the world, was completed in December 2010. The purpose of the detector is to investigate high energy neutrinos, search for dark matter, observe supernovae explosions, and search for exotic particles such as magnetic monopoles.

- ANTARES (telescope). (Toulon, France). A Neutrino detector 2.5 km under the Mediterranean Sea off the coast of Toulon, France. Designed to locate and observe neutrino flux in the direction of the southern hemisphere.

- XENONnT, the upgrade of XENON1T, is a dark matter direct search experiment located at the Gran Sasso National Laboratories and will be sensitive to WIMPs with SI cross section of $10^{-48}$ cm$^2$.

- BOREXINO, a real-time detector, installed at Laboratori Nazionali del Gran Sasso, designed to detect neutrinos from the Sun with an organic liquid scintillator target.

- Pierre Auger Observatory (Malargüe, Argentina). Detects and investigates high energy cosmic rays using two techniques. One is to study the particles interactions with water placed in surface detector tanks. The other technique is to track the development of air showers through observation of ultraviolet light emitted high in the Earth's atmosphere.

- CERN Axion Solar Telescope (CERN, Switzerland). Searches for axions originating from the Sun.

- NESTOR Project (Pylos, Greece). The target of the international collaboration is the deployment of a neutrino telescope on the sea floor off of Pylos, Greece.

- Kamioka Observatory is a neutrino and gravitational waves laboratory located underground in the Mozumi Mine near the Kamioka section of the city of Hida in Gifu Prefecture, Japan.

- Laboratori Nazionali del Gran Sasso is a laboratory that hosts experiments that require a low noise background environment. Located within the Gran Sasso mountain, near L'Aquila (Italy). Its experimental halls are covered by 1400m of rock, which protects experiments from cosmic rays.

- SNOLAB.

- Aspera European Astroparticle network Started in July 2006 and is responsible for coordinating and funding national research efforts in Astroparticle Physics.

- Telescope Array Project (Delta, Utah) An experiment for the detection of ultra high energy cosmic rays (UHECRs) using a ground array and fluorescence techniques in the desert of west Utah.

## Nuclear Astrophysics

Nuclear astrophysics is an interdisciplinary branch of physics involving close collaboration among researchers in various subfields of nuclear physics and astrophysics: notably stellar modeling; measurement and theoretical estimation of nuclear reaction rates; physical cosmology and cosmochemistry; gamma ray, optical and X-ray astronomy; and extending our knowledge about nuclear lifetimes and masses. In general terms, nuclear astrophysics aims *to understand the origin of the chemical elements and the energy generation in stars.*

The basic principles for explaining the origin of elements and energy generation in stars appear in the theory of nucleosynthesis, which came together in the late 1950s in seminal works by Burbidge, Burbidge, Fowler, and Hoyle, and by Cameron. Fowler is largely credited with initiating collaboration between astronomers, astrophysicists, and experimental nuclear physicists that we now know as nuclear astrophysics (for which he won the 1983 Nobel Prize).

The basic tenets of nuclear astrophysics are that only isotopes of hydrogen and helium (and traces of lithium, beryllium, and boron) can be formed in a homogeneous Big Bang model, while all other elements are formed in stars. Conversion of nuclear mass to radiative energy (per Einstein's famous mass-energy relation) is what allows stars to shine for up to billions of years. Many notable physicists of the 19th century such as Mayer, Waterson, von Helmholtz, and Lord Kelvin, postulated that the Sun radiates thermal energy by converting gravitational potential energy into heat. Under such a model, its lifetime can be calculated relatively easily using the virial theorem — around 19 million years, which was inconsistent with the interpretation of geological records and the (then new) theory of biological evolution.

A back-of-the-envelope calculation indicates that if the Sun consisted entirely of a fossil fuel like coal (a source of energy familiar to many), considering the rate of thermal energy emission, its lifetime would be merely four or five thousand years, which is not even consistent with records of human civilization. Though now discredited, this hypothesis that the Sun's primary energy source is gravitational contraction was reasonable before the advent of modern physics; radioactivity itself was not discovered by Becquerel until 1895. Besides the prerequisite knowledge of the atomic nucleus, a proper understanding of stellar energy is not possible without the theories of relativity and quantum mechanics.

After Aston demonstrated that the mass of helium is less than four times that of the proton, Eddington proposed that, through an unknown process in the Sun's core, hydrogen is transmuted into helium, liberating energy. Twenty years later, Bethe and von Weizsäcker independently derived the CN cycle, the first known nuclear reaction that accomplishes this transmutation. However, the Sun's primary energy source is now understood to be proton–proton chain reactions, occurring at much lower energies and much more slowly than catalytic hydrogen fusion.

The interval between Eddington's proposal and derivation of the CN cycle can mainly be attributed to an incomplete understanding of nuclear structure. A proper understanding of nucleosynthetic processes only came when Chadwick discovered the neutron in 1932 and beta decay theory developed. Nuclear physics gives a picture of the Sun's energy source producing a lifetime consistent with the age of the Solar System derived from meteoritic abundances of lead and uranium isotopes — about 4.5 billion years. The mass of stars like the Sun allow core hydrogen burning on the main sequence of the Hertzsprung-Russell diagram via pp-chains for about 9 billion years. This primarily is determined by extremely slow production of deuterium,

$$\,^1_1H + \,^1_1H \rightarrow \,^2_1D + e^+ + v_e + 0.42 \text{ MeV}$$

which is governed by the nuclear weak force.

Abundances of the chemical elements in the Solar System. Hydrogen and helium are most common, residuals within the paradigm of the Big Bang. The next three elements (Li, Be, B) are rare because they are poorly synthesized in the Big Bang and also in stars. The two general trends in the remaining stellar-produced elements are: (1) an alternation of abundance of elements according to whether they have even or odd atomic numbers, and (2) a general decrease in abundance, as elements become heavier. Within this trend is a peak at abundances of iron and nickel, which is especially visible on a logarithmic graph spanning fewer powers of ten, say between logA=2 (A=100) and logA=6 (A=1,000,000).

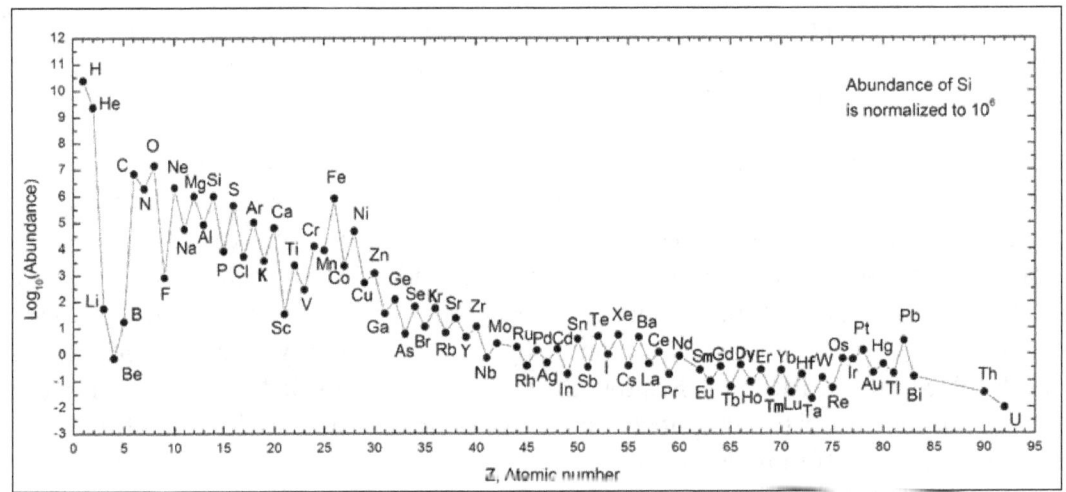

## Predictions

Stellar nucleosynthesis theory estimates chemical abundances consistent with those observed in the Solar System and galaxy, whose distribution spans twelve orders of magnitude (one trillion). While impressive, these data were used to formulate the theory, but a scientific theory must be predictive to have merit. The theory has been well-tested by observation and experiment since first formulated.

The theory predicts technetium (the lightest chemical element without stable isotopes) in stars, galactic gamma-emitters (such as $^{26}$Al and $^{44}$Ti), and observation of solar neutrinos and from supernova 1987a. These observations have far-reaching implications. $^{26}$Al has a lifetime a bit less than one million years, which is very short on a galactic timescale, proving that nucleosynthesis is an ongoing process even in our own time. Work that led to discovery of neutrino oscillation (implying a non-zero mass for the neutrino absent in the Standard Model of particle physics) was motivated by a solar neutrino flux about three times lower than expected — a long-standing concern in the nuclear astrophysics community colloquially known as the Solar neutrino problem. The observable neutrino flux from nuclear reactors is much larger than that of the Sun, so Davis and others were primarily motivated to look for solar neutrinos for astronomical reasons.

# Atomic and Molecular Astrophysics

Within a few million years the light from bright stars will have boiled away this molecular cloud of gas and dust. The cloud has broken off from the Carina Nebula. Newly formed stars are visible nearby, their images reddened by blue light being preferentially scattered by the pervasive dust. This image spans about two light years and was taken by the orbiting Hubble Space Telescope in 1999.

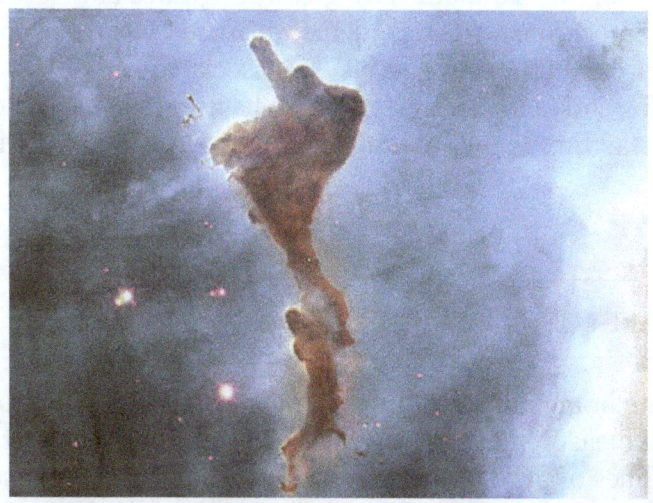

Atomic astrophysics is concerned with performing atomic physics calculations that will be useful to astronomers and using atomic data to interpret astronomical observations. Atomic physics plays a key role in astrophysics as astronomers' only information about a particular object comes through the light that it emits, and this light arises through atomic transitions.

Molecular astrophysics, developed into a rigorous field of investigation by theoretical astrochemist Alexander Dalgarno beginning in 1967, concerns the study of emission from molecules in space. There are 110 currently known interstellar molecules. These molecules have large numbers of observable transitions. Lines may also be observed in absorption—for example the highly redshifted lines seen against the gravitationally lensed quasar PKS1830-211. High energy radiation, such as ultraviolet light, can break the molecular bonds which hold atoms in molecules. In general then, molecules are found in cool astrophysical environments. The most massive objects in our galaxy are giant clouds of molecules and dust known as giant molecular clouds. In these clouds, and smaller versions of them, stars and planets are formed. One of the primary fields of study of molecular astrophysics is star and planet formation. Molecules may be found in many environments, however, from stellar atmospheres to those of planetary satellites. Most of these locations are relatively cool, and molecular emission is most easily studied via photons emitted when the molecules make transitions between low rotational energy states. One molecule, composed of the abundant carbon and oxygen atoms, and very stable against dissociation into atoms, is carbon monoxide (CO). The wavelength of the photon emitted when the CO molecule falls from its lowest excited state to its zero energy, or ground, state is 2.6mm, or 115 gigahertz. This frequency is a thousand times higher than typical FM radio frequencies. At these high frequencies, molecules in the Earth's atmosphere can block transmissions from space, and telescopes must be located in dry (water is an important atmospheric blocker), high sites. Radio telescopes must have very accurate surfaces to produce high fidelity images.

On February 21, 2014, NASA announced a greatly upgraded database for tracking polycyclic aromatic hydrocarbons (PAHs) in the universe. According to scientists, more than 20% of the carbon in the universe may be associated with PAHs, possible starting materials for the formation of life. PAHs seem to have been formed shortly after the Big Bang, are widespread throughout the universe, and are associated with new stars and exoplanets.

## Solar Physics

Solar physics is the branch of astrophysics that specializes in the study of the Sun. It deals with detailed measurements that are possible only for our closest star. It intersects with many disciplines of pure physics, astrophysics, and computer science, including fluid dynamics, plasma physics including magnetohydrodynamics, seismology, particle physics, atomic physics, nuclear physics, stellar evolution, space physics, spectroscopy, radiative transfer, applied optics, signal processing, computer vision, computational physics, stellar physics and solar astronomy.

Because the Sun is uniquely situated for close-range observing (other stars cannot be resolved with anything like the spatial or temporal resolution that the Sun can), there is a split between the related discipline of observational astrophysics (of distant stars) and observational solar physics.

The study of solar physics is also important as it provides a "physical laboratory" for the study of plasma physics.

### Ancient Times

Babylonians were keeping a record of solar eclipses, with the oldest record originating from the ancient city of Ugarit, in modern-day Syria. This record dates to about 1300 BC. Ancient Chinese astronomers were also observing solar phenomena (such as solar eclipses and visible sunspots) with the purpose of keeping track of calendars, which were based on lunar and solar cycles. Unfortunately, records kept before 720 BC are very vague and offer no useful information. However, after 720 BC, 37 solar eclipses were noted over the course of 240 years.

## Medieval Times

Astronomical knowledge flourished in the Islamic world during medieval times. Many observatories were built in cities from Damascus to Baghdad, where detailed astronomical observations were taken. Particularly, a few solar parameters were measured and detailed observations of the Sun were taken. Solar observations were taken with the purpose of navigation, but mostly for timekeeping. Islam requires its followers to pray five times a day, at specific position of the Sun in the sky. As such, accurate observations of the Sun and its trajectory on the sky were needed. In the late 10th century, Iranian astronomer Abu-Mahmud Khojandi built a massive observatory near Tehran. There, he took accurate measurements of a series of meridian transits of the Sun, which he later used to calculate the obliquity of the ecliptic. Following the fall of the Western Roman Empire, Western Europe was cut from all sources of ancient scientific knowledge, especially those written in Greek. This, plus de-urbanisation and diseases such as the Black Death led to a decline in scientific knowledge in Medieval Europe, especially in the early Middle Ages. During this period, observations of the Sun were taken either in relation to the zodiac, or to assist in building places of worship such as churches and cathedrals.

## Renaissance Period

In astronomy, the renaissance period started with the work of Nicolaus Copernicus. He proposed that planets revolve around the Sun and not around the Earth, as it was believed at the time. This model is known as the heliocentric model. His work was later expanded by Johannes Kepler and Galileo Galilei. Particularly, Galilei used his new telescope to look at the Sun. In 1610, he discovered sunspots on its surface. In the autumn of 1611, Johannes Fabricius wrote the first book on sunspots, *De Maculis in Sole Observatis* ("On the spots observed in the Sun").

## Modern Times

Modern day solar physics is focused towards understanding the many phenomena observed with the help of modern telescopes and satellites. Of particular interest are the structure of the solar photosphere, the coronal heat problem and sunspots.

## Research in Solar Physics

The Solar Physics Division of the American Astronomical Society boasts 555 members, compared to several thousand in the parent organization.

A major thrust of current effort in the field of solar physics is integrated understanding of the entire Solar System including the Sun and its effects throughout interplanetary space within the heliosphere and on planets and planetary atmospheres. Studies of phenomena that affect multiple systems in the heliosphere, or that are considered to fit within a heliospheric context, are called heliophysics, a new coinage that entered usage in the early years of the current millennium.

## Space based

### SDO

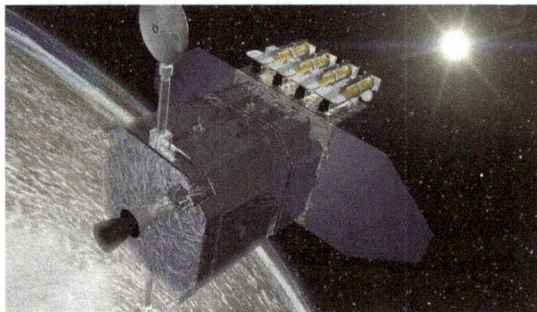

The SDO satellite.

The Solar Dynamics Observatory (SDO) was launched by NASA in February 2010 from Cape Canaveral. The main goals of the mission are understanding how solar activity arises and how it affects life on Earth by determining how the Sun's magnetic field is generated and structured and how the stored magnetic energy is converted and released into space.

### SOHO

Image of SOHO spacecraft.

The Solar and Heliospheric Observatory, SOHO, is a joint project between NASA and ESA that was launched in December 1995. It was launched to probe the interior of the Sun, make observations of the solar wind and phenomena associated with it and investigate the outer layers of the Sun.

## HINODE

A publicly funded mission led by the Japanese Aerospace Exploration Agency, the HINODE satellite, launched in 2006, consists of a coordinated set of optical, extreme ultraviolet and X-ray instruments. These investigate the interaction between the solar corona and the Sun's magnetic field.

## Ground based

## ATST

The Advanced Technology Solar Telescope (ATST) is a solar telescope facility that is under construction in Maui. Twenty-two institutions are collaborating on the ATST project, with the main funding agency being the National Science Foundation.

## SSO

Sunspot Solar Observatory (SSO) operates the Richard B. Dunn Solar Telescope (DST) on behalf of the NSF.

## Big Bear

The Big Bear Solar Observatory in California houses several telescopes including the New Solar Telescope(NTS) which is a 1.6 meter, clear-aperture, off-axis Gregorian telescope. The NTS saw first light in December 2008. Until the ATST comes on line, the NTS remains the largest solar telescope in the world. The Big Bear Observatory is one of several facilities operated by the Center for Solar-Terrestrial Research at New Jersey Institute of Technology (NJIT).

## Other

## EUNIS

The Extreme Ultraviolet Normal Incidence Spectrograph (EUNIS)is a two channel imaging spectrograph that first flew in 2006. It observes the solar corona with high spectral resolution. So far, it has provided information on the nature of coronal bright points, cool transients and coronal loop arcades. Data from it also helped calibrating SOHO and a few other telescopes.

# Astrophysical Fluid Dynamics

Astrophysical fluid dynamics is a modern branch of astronomy involving fluid mechanics which deals with the motion of fluids, like the gases which the stars are made up of or any fluid which is found in outer space. The subject covers the fundamentals of mechanics of fluids using various equations, ranging from the continuity equation, Navier Stokes to Euler's equations of collisional fluids and the like. It is an extensive study of the physical realms of the astral bodies and their movements in space. A thorough understanding of this subject requires detailed knowledge of the equations governing fluid mechanics. Most of the applications of astrophysical fluid dynamics include dynamics of stellar systems, accretion disks, Astrophysical jets, Newtonian fluids, and the fluid dynamics of galaxies.

Astrophysical fluid dynamics deals with the application of fluid dynamics and its equations in the movement of the fluids in space. The applications are entirely different from what we usually study as all of this happens in vacuum with zero or lesser gravity.

Most of the Interstellar Medium is not at rest, but is in supersonic motion under the action of supernova explosions, stellar winds and radiation fields and the time dependent gravitational field due to spiral density waves in the stellar disc of the galaxy. Since supersonic motions almost always involve shock waves, these play a crucial role. The galaxy also contains a dynamically significant magnetic field which means that the dynamics is governed by the equations of compressible magnetohydrodynamics.

In many cases the electrical conductivity is large enough for the ideal magnetohydrodynamics to be a good approximation, but this is not true in star forming regions where the gas density is high and the degree of ionization is low.

One of the most interesting problems is that of star formation. It is known that stars form out of the Interstellar Medium and that this mostly occurs in Giant Molecular Clouds such as the Rosette Nebula for example. It has been known for a long time that an interstellar cloud can collapse due to its self-gravity if it is large enough, but in the ordinary interstellar medium, this can only happen if the cloud has a mass of several thousand solar masses - much larger than that of any star. There must therefore be some process that fragments the cloud into smaller high density clouds whose masses are in the same range as that of stars. Self-gravity cannot do this, but it turns out that there are processes that do this if the magnetic pressure is much larger than the thermal pressure, as it is in Giant Molecular Clouds. These processes rely on the interaction of magnetohydrodynamic waves with a thermal instability. A magnetohydrodynamic wave in a medium in which the magnetic pressure is much larger than the thermal pressure can produce dense regions, but they cannot by themselves make the density high enough for self-gravity to act. However, the gas in star forming regions is heated by cosmic rays and is cooled by radiative processes. The net result is that gas in a thermal equilibrium state in which heating balances cooling can exist in three different phases at the same pressure: a warm phase with a low density, an unstable phase with intermediate density and a cold phase at low temperature. An increase in pressure, due to a supernova or a spiral density wave can flip the gas from the warm phase into the unstable phase and a Magnetohydrodynamic wave can then produce dense fragments in the cold phase whose self-gravity is strong enough for them to collapse to form stars.

In this process, we can study the dynamics of the cosmic gas and understand the formation of stars. This is just one example. Even Magnetohydrodynamics has its basis on the fundamentals of astrophysical fluid dynamics.

## Basic Concepts

## Concepts of Fluid Dynamics

The equations of Fluid Dynamics are tools in developing an understanding of the phenomena in Astrophysical Fluid Dynamics.The important equations with their applications are as mentioned below:

## Conservation of Mass

The continuity equation applies the principle of conservation of mass to fluid flow. Consider a fluid flowing through a fixed volume tank having one inlet and one outlet as shown below:

If the flow is steady i.e. no accumulation of fluid within the tank, then the rate of fluid flow at entry must be equal to the rate of fluid flow at exit for mass conservation. If, at entry (or exit) having a cross-sectional area A (m²), a fluid parcel travels a distance dL in time dt, then the volume flow rate (V, m³/s) is given by: $V = (A . dL)/\Delta t$.

But since dL/$\Delta$t is the fluid velocity (v, m/s) we can write: $Q = V x A$.

The mass flow rate (m, kg/s) is given by the product of density and volume flow rate,

$$i.e \ m = \rho.Q = \rho .V.A$$

Between two points in flowing fluid for mass conservation we can write: $m_1 = m_2$,

$$Or \ \rho_1 \ V_1 \ A_1 = \rho_2 \ V_2 \ A_2$$

If the fluid is *incompressible* i.e. $\rho_1 = \rho_2$ then:

$$V_1 A_1 = V_2 A_2.$$

But, We shall apply this theorem for Astrophysicsical Fluid Dynamics in supersonic Flow regime which will require us to consider a Compressible flow condition where density is not constant.

An application for fluid dynamics in astrophysics is the Neutron stars, which are ancient remnants of stars that have reached the end of their evolutionary journey through space and time.

These interesting objects are born from once-large stars that grew to four to eight times the size of our own sun before exploding in catastrophic supernovae. After such an explosion blows a star's outer layers into space, the core remains—but it no longer produces nuclear fusion. With no outward pressure from fusion to counterbalance gravity's inward pull, the star condenses and collapses in upon itself.

Despite their small diameters—about 12.5 miles (20 kilometers)—neutron stars boast nearly 1.5 times the mass of our sun, and are thus incredibly dense. Just a sugar cube of neutron star matter would weigh about one hundred million tons on Earth.

A neutron star's almost incomprehensible density causes protons and electrons to combine into neutrons—the process that gives such stars their name. The composition of their cores is unknown, but they may consist of a neutron superfluid or some unknown state of matter.

Neutron stars pack an extremely strong gravitational pull, much greater than Earth's. This gravitational strength is particularly impressive because of the stars' small size.

When they are formed, neutron stars rotate in space. As they compress and shrink, this spinning speeds up because of the conservation of angular momentum—the same principle that causes a spinning skater to speed up when she pulls in her arms.

These stars gradually slow down over the eons, but those bodies that are still spinning rapidly may emit radiation that from Earth appears to blink on and off as the star spins, like the beam of light from a turning lighthouse. This "pulsing" appearance gives some neutron stars the name pulsars.

After spinning for several million years pulsars are drained of their energy and become normal neutron stars. Few of the known existing neutron stars are pulsars. Only about 1,000 pulsars are known to exist, though there may be hundreds of millions of old neutron stars in the galaxy.

The staggering pressures that exist at the core of neutron stars may be like those that existed at the time of the big bang, but these states cannot be simulated on Earth.

## EMG (Estakhr's Material Geodesic) Equations

It seems EMG Equations plays the most important role in this new branch of Astronomy. This equation was introduced for the first time by American Physical Society in 2013. Estakhr's Material-Geodesic equations is developed model of Navier-Stokes equations in an umbrella term, It is relativistic version of NS-equations, And that is why it is so important.

## Stellar Dynamics

Stellar dynamics is the branch of astrophysics which describes in a statistical way the collective motions of stars subject to their mutual gravity. The essential difference from celestial mechanics is that each star contributes more or less equally to the total gravitational field, whereas in celestial mechanics the pull of a massive body dominates any satellite orbits.

Historically, the methods utilized in stellar dynamics originated from the fields of both classical mechanics and statistical mechanics. In essence, the fundamental problem of stellar dynamics is the N-body problem, where the N members refer to the members of a given stellar system. Given the large number of objects in a stellar system, stellar dynamics is usually concerned with the more global, statistical properties of several orbits rather than with the specific data on the positions and velocities of individual orbits.

The motions of stars in a galaxy or in a globular cluster are principally determined by the average distribution of the other, distant stars. Stellar encounters involve processes such as relaxation, mass segregation, tidal forces, and dynamical friction that influence the trajectories of the system's members.

Stellar dynamics also has connections to the field of plasma physics. The two fields underwent significant development during a similar time period in the early 20th century, and both borrow mathematical formalism originally developed in the field of fluid mechanics.

### Key Concepts

Stellar dynamics involves determining the gravitational potential of a substantial number of stars. The stars can be modeled as point masses whose orbits are determined by the combined interactions with each other. Typically, these point masses represent stars in a variety of clusters or galaxies, such as a Galaxy cluster, or a Globular cluster. From Newton's second law an equation describing the interactions of an isolated stellar system can be written down as,

$$m_i \frac{d\mathbf{r_i}}{dt} = \sum_{\substack{i=1 \\ i \neq j}}^{N} \frac{Gm_i m_j \left(\mathbf{r}_i - \mathbf{r}_j\right)}{\left\|\mathbf{r}_i - \mathbf{r}_j\right\|^3}$$

which is simply a formulation of the N-body problem. For an N-body system, any individual member, $m_i$ is influenced by the gravitational potentials of the remaining $m_j$ members. In practice, it is not feasible to calculate the system's gravitational potential by adding all of the point-mass potentials in the system, so stellar dynamicists develop potential models that can accurately model the system while remaining computationally inexpensive. The gravitational potential, $\Phi$, of a system is related to the gravitational field, $\vec{g}$ by:

$$\vec{g} = -\nabla\Phi$$

whereas the mass density, $\rho$, is related to the potential via Poisson's equation:

$$\nabla^2\Phi = 4\pi G\rho$$

## Gravitational Encounters and Relaxation

Stars in a stellar system will influence each other's trajectories due to strong and weak gravitational encounters. An encounter between two stars is defined to be strong if the change in potential energy between the two is greater than or equal to their initial kinetic energy. Strong encounters are rare, and they are typically only considered important in dense stellar systems, such as the cores of globular clusters. Weak encounters have a more profound effect on the evolution of a stellar system over the course of many orbits. The effects of gravitational encounters can be studied with the concept of relaxation time.

A simple example illustrating relaxation is two-body relaxation, where a star's orbit is altered due to the gravitational interaction with another star. Initially, the subject star travels along an orbit with initial velocity, $v$, that is perpendicular to the impact parameter, the distance of closest approach, to the field star whose gravitational field will affect the original orbit. Using Newton's laws, the change in the subject star's velocity, $\delta v$, is approximately equal to the acceleration at the impact parameter, multiplied by the time duration of the acceleration. The relaxation time can be thought as the time it takes for $\delta v$ to equal $v$, or the time it takes for the small deviations in velocity to equal the star's initial velocity. The relaxation time for a stellar system of $N$ objects is approximately equal to:

$$t_{\text{relax}} \simeq \frac{0.1N}{\ln N} t_{\text{cross}}$$

where $t_{\text{cross}}$ is known as the crossing time, the time it takes for a star to travel across the galaxy once.

The relaxation time identifies collisionless vs. collisional stellar systems. Dynamics on timescales less than the relaxation time are defined to be collisionless. They are also identified as systems where subject stars interact with a smooth gravitational potential as opposed to the sum of point-mass potentials. The accumulated effects of two-body relaxation in a galaxy can lead to what is known as mass segregation, where more massive stars gather near the center of clusters, while the less massive ones are pushed towards the outer parts of the cluster.

## Connections to Statistical Mechanics and Plasma Physics

The statistical nature of stellar dynamics originates from the application of the kinetic theory of gases to stellar systems by physicists such as James Jeans in the early 20th century. The Jeans equations, which describe the time evolution of a system of stars in a gravitational field, are analogous to Euler's equations for an ideal fluid, and were derived from the collisionless Boltzmann equation. This was originally developed by Ludwig Boltzmann to describe the non-equilibrium behavior of a thermodynamic system. Similarly to statistical mechanics, stellar dynamics make use of distribution functions that encapsulate the information of a stellar system in a probabilistic manner. The single particle phase-space distribution function, $f(\mathbf{x}, \mathbf{v}, t)$, is defined in a way such that,

$$f(\mathbf{x}, \mathbf{v}, t) \mathrm{d}\mathbf{x}\, \mathrm{d}\mathbf{v}$$

represents the probability of finding a given star with position $\mathbf{x}$ around a differential volume $\mathrm{d}\mathbf{x}$ and velocity $\mathbf{v}$ around a differential volume $\mathrm{d}\mathbf{v}$. The distribution is function is normalized such that integrating it over all positions and velocities will equal unity. For collisional systems, Liouville's theorem is applied to study the microstate of a stellar system, and is also commonly used to study the different statistical ensembles of statistical mechanics.

In plasma physics, the collisionless Boltzmann equation is referred to as the Vlasov equation, which is used to study the time evolution of a plasma's distribution function. Whereas Jeans applied the collisionless Boltzmann equation, along with Poisson's equation, to a system of stars interacting via the long range force of gravity, Anatoly Vlasov applied Boltzmann's equation with Maxwell's equations to a system of particles interacting via the Coulomb Force. Both approaches separate themselves from the kinetic theory of gases by introducing long-range forces to study the long term evolution of a many particle system. In addition to the Vlasov equation, the concept of Landau damping in plasmas was applied to gravitational systems by Donald Lynden-Bell to describe the effects of damping in spherical stellar systems.

## Applications

Stellar dynamics is primarily used to study the mass distributions within stellar systems and galaxies. Early examples of applying stellar dynamics to clusters include Albert Einstein's 1921 paper applying the virial theorem to spherical star clusters and Fritz Zwicky's 1933 paper applying the virial theorem specifically to the Coma Cluster, which was one of the original harbingers of the idea of dark matter in the universe. The Jeans equations have been used to understand different observational data of stellar motions in the Milky Way galaxy. For example, Jan Oort utilized the Jeans equations to determine the average matter density in the vicinity of the solar neighborhood, whereas the concept of asymmetric drift came from studying the Jeans equations in cylindrical coordinates.

Stellar dynamics also provides insight into the structure of galaxy formation and evolution. Dynamical models and observations are used to study the triaxial structure of elliptical galaxies and suggest that prominent spiral galaxies are created from galaxy mergers. Stellar dynamical models are also used to study the evolution of active galactic nuclei and their black holes, as well as to estimate the mass distribution of dark matter in galaxies.

## References

- De Angelis, Alessandro; Pimenta, Mario (2018). Introduction to particle and astroparticle physics (multimessenger astronomy and its particle physics foundations). Springer. Doi:10.1007/978-3-319-78181-5. ISBN 978-3-319-78181-5

- Hoover, Rachel (February 21, 2014). "Need to Track Organic Nano-Particles Across the Universe? NASA's Got an App for That". NASA. Retrieved February 22, 2014

- Computational-astrophysics: scholarpedia.org, Retrieved 31 July 2019

- "Estakhr's Relativistic Decomposition of Four-Velocity Vector Field of Big Bang (Big Bang's Turbulence)". APS. American Physical Society. Retrieved 2016-09-22

- Binney, James; Tremaine, Scott (2008). Galactic Dynamics. Princeton: Princeton University Press. Pp. 35, 63, 65, 698. ISBN 978-0-691-13027-9

# Basic Concepts in Astrophysics

**3**

- **Astronomical Coordinate Systems**
- **Astronomical Unit**
- **Astronomical Unit of Mass**
- **Big Bang Theory**
- **Cosmic Microwave Background**
- **Luminosity**
- **Dark Matter**
- **Dark Energy**
- **Accretion**
- **Optical Depth**

Some of the fundamental concepts within the field of astrophysics are Big Bang theory, dark matter, dark energy, optical depth, accretion, cosmic microwave background and astronomical coordinate systems. The topics elaborated in this chapter such as will help in gaining a better perspective about these concepts of astrophysics.

## Astronomical Coordinate Systems

### Positions and Coordinate Systems

One of the basic needs of astronomy, as well as other physical sciences, is to give reasonable descriptions for the positions of objects relative to each other. Scientifically, this is done in mathematical language, by properly assigning numbers to each position in space; these numbers are called *coordinates* and the system defined by this procedure a *coordinate system.*

The coordinate systems considered here are all based at one reference point in space with respect to which the positions are measured, the *origin* of the reference frame (typically, the location of the observer, or the center of Earth, the Sun, or the Milky Way Galaxy). Any location in space is then described by the "radius vector" or "arrow" between the origin and the location, namely by the *distance* (length of the vector) and its *direction*. The direction is given by the straight half line from the origin through the location (to infinity).

A reference plane containing the origin is fixed, or equivalently the axis through the origin and perpendicular to it (typically, an "equatorial" plane and a "polar" axis); elementarily, each of these uniquely determines the other. One can assign an orientation to the polar axis from "negative" to "positive", or "south" to "north", and simultaneously to the equatorial plane by assigning a positive sense of rotation to the equatorial plane; these orientations are, by convention, usually combined by the right hand rule: If the thumb of the right hand point to the positive (north) polar axis, the fingers show in the positive direction of rotation (and vice versa, so that a physical rotation defines a north direction).

The reference plane or the reference axis define the set of planes which contain the origin and are perpendicular to the "equatorial" reference plane (or equivalently, contain the "polar" reference axis); each direction in space then lies precisely in one of these "meridional" planes (or half planes, if the reference axis is taken to divide each plane into halfs), with the exception of the (positive and negative) polar axis which lies in all of them by definition.

The first angle used to characterize a direction, typically the "latitude", is taken between the direction and the reference plane, within the "meridional" plane. For the second angle, it is required to select and fix one of the "meridional" half planes as zero, from which the angle (of "longitude") is measured to the "meridional" half plane containing our direction.

Note that this selection of angles to characterize a direction in a given reference frame is chosen by convention, which is especially common in astronomy and geography, and which is used in the following here, as well as in most astronomical databases. Other, equivalent, conventions are possible, e.g. physicists often use instead of the "latitude" angle to the reference plane, the angle between the direction and the "positive" or "north" polar axis (called "co-latitude"; co-latitde = 90 deg - latitude). It depends on taste at last what the reader likes to use, but here we will stay as close to standard astronomical convention as possible. In order to minimize the requirement of case-to-case enumeration of conventions, we also recommend the reader to do the same.

## Positions on Earth

On Earth, positions are usually given by two angles, the longitude and the latitude coordinates on the surface of Earth, and the elevation of the location "above (or below) sea level"; sea level is defined here by a certain value of the Earth's gravitational potential (equipotential surface).

The natural reference plane here is that of the Earth's equator, and the natural reference axis is the rotational, polar axis which cuts the Earth's surface at the planet's North and South pole. The circles along Earth's surface which are parallel to the equator are the *latitude circles*, where the angle at the planet's center is constant for all points on these circles. Half circles from pole to pole, which are all perpendicular to the equatorialplane, are called *meridians*. One of the meridians, in practice that through the Greenwich Observatory near London, England, is taken as reference meridian, or *Null meridian*. *Geographical Longitude* is measured as the angle between this and the meridian under consideration (or more precisely, between the half planes containing them); it is of course the same for all points of the meridian.

Because Earth is not exactly circular, but slightly flattened, its surface (defined by the ocean surface, or the corresponding gravitational potential) forms a specific figure, the so-called *Geoid*, which is very similar to a slightly oblate *spheroid* (the *reference ellipsoid*). This is the reason why there are two common but different definitions of *latitude* on Earth:

- *Geocentric Latitude*, measured as angle at the Earth's center, between the equatorial plane and the direction to the surface point under consideration, and

- *Geographical Latitude*, measured on the surface between the parallel plane to the equatorial plane and the line orthogonal to the surface, the *local vertical* or plumb line, which may be measured by the direction of gravitional force (e.g., plumb).

The absolute value of the geocentric latitude is always smaller or equal (at poles and equator) to the absolute value of the geographical latitude.

Taking any of the meridional planes, the meridian has the approximate shape of a half ellipse. The major half axis represents the equatorial radius of the planet, while the minor axis is the polar (and thus the rotational) half axis, which is about 1/298 shorter than the equatorial radius. More precisely:

- Equatorial radius: a = 6378.140 km.

- Polar radius: b = 6356.755 km.

- Flattening/Oblateness: f = 1/298.253.

One can calculate the difference in the two latitudes for every point on the meridian.

Denoting the geographical latitude with B, the geocentrical with B', one obtains after some calculation:

$$\tan B' = (b/a)^2 * \tan B$$

The maximal difference occurs for B = 45 deg, and amounts to 11.5 arc minutes.

In the following, we always deal with geographic latitude unless otherwise mentioned.

## The Celestial Sphere

Compared to the size of Earth, all celestial objects (with the exception of some satellites and meteorites in Earth's atmosphere) are far away. Viewing them, they look all at far distance, which can not be distinguished easily, so that they look as being positioned on a far-away sphere.

Thus each observer can look at the skies as being manifested on the interior of a big sphere, the so-called *celestial sphere*. Then each direction away from the observer will intersect the celestial sphere in one unique point, and positions of stars and other celestial objects can be measured in angular coordinates (similar to longitude and latitude on Earth) on this virtual sphere. This can be done without knowing the actual distances of the stars. Moreover, any plane through the origin cuts the sphere in a great circle. Examples for celestial coordinate systems are treated below.

In times up to Copernicus, people believed that there is actually a solid sphere to which the stars beyond the solar system are fixed: This idea was overcome when it was realized that stars are sunlike bodies, in the time of Newton and Halley. Today, the celestial sphere is only a virtual construct to make our understanding of positional astronomy easier.

## The Horizon System

The horizon system is defined locally for each observer, or site, on Earth (or another celestial body). Its origin is the observer's location, its reference axis is the *local vertical* or plumb line (defined e.g. by the local gravitational field), and its reference plane is the *apparent horizon* or simply *horizon* perpendicular to it at the observer's location. The direction directly, or vertically, above the observer, or its intersection point with the virtual celestial sphere, is called *zenith*, the opposite direction or point, vertically below the observer, is called *nadir*.

Through any direction, or point on the celestial sphere, e.g. the position of a star, a unique [half] plane (or great [half] circle) perpendicular to the horizon can be found; this is called *vertical circle*; all vertical [half] circles contain (and intersect in) both the zenith and the nadir. Within the plane of its verticle circle, the position under consideration can be characterized by the angle to the horizon, called *altitude* a. Alternatively and equivalently, one could take the angle between the direction and the zenith, the *zenith distance* z, which is related to the altitude by the relation: z = 90 deg - a. All objects *above* the horizon have positive altitudes (or zenith distances smaller than 90 deg). The horizon itself can be defined, or recovered, as the set of all points for which a = 0 deg (or z = 90 deg).

In contrast to the apparent horizon which defines coordinates of objects as the observer perceives them, the *true horizon* is defined by the plane parallel to the apparent horizon, but through the center of Earth. The angle between the position of an object and the true horizon is referred to as *true altitude*. For nearby objects such as the Moon, the measured position can vary notably between these two reference systems (up to 1 deg for the Moon). Also, the apparent altitudes are subject to the effect of refraction by Earth's atmosphere.

The second coordinate of a position in the horizon system is defined by the point where the verticle circle of the position cuts the horizon. It is called *azimuth* A and, in astronomy and on the Northern hemisphere (the present author does not know the southern standards for this thread), is the angle from the south point (or direction) taken to the west, north, and east to the foot point of the vertical circle on the horizon, thus running from 0 to 360 deg. In geodesy, the north direction is often taken as zero point (this angle is sometimes called *bearing* and is given by A +/- 180 deg). Note that these conventions are not always uniquely used so that it may be advisable to clear up which conventions are used (e.g., by saying A is taken to the West).

Taking the astronomical standard, the *south, west, north*, and *east* points on the horizon are defined by A = 0 deg, 90 deg, 180 deg, and 270 deg, respectively. The vertical circle passing through the south and north point (as well as zenith and nadir) is called *local meridian*; the one perpendicular to it through west point, zenith, east point and nadir is called *prime vertical*. The local meridian coincides with the projection of the geographical meridian of the observer's location to the sky (celestial sphere) from Earth's center.

The terms introduced here are helpful in understanding the effects of Earth's rotation.

## The Equatorial Coordinate System

The main disadvantage of the horizon system is the steady change of coordinates for a given astronomical object as Earth rotates during the course of the day. This can be removed by using a coordinate system which is fixed at the stars (or the celestial sphere). The most frequently used such system is the *equatorial coordinate system* which is still related to planet Earth and thus convenient for observers.

In principle, the celestial coordinate system can be introduced in the simplest way by projecting Earth's geocentric coordinates to the sky at a certain moment of time (actually, each time when star time is 0:00 at Greenwich or anywhere on the Zero meridian on Earth, which occurs once each siderial day). These coordinates are then left fixed at the celestial sphere, while Earth will rotate away below them.

Practically, projecting Earth's equator and poles to the celestial sphere by imagining straight half lines from the Earth's center produces the *celestial equator* as well as the *north* and the *south celestial pole*. Great circles through the celestial poles are always perpendicular to the celestial equator and called *hour circles* for reasons explained below:

The first coordinate in the equatorial system, corresponding to the latitude, is called Declination (Dec), and is the angle between the position of an object and the celestial equator (measured along the hour circle). Alternatively, sometimes the *polar distance* (PD) is used, which is given by PD = 90 deg - Dec; the most prominent reference known to the present author using PD instead of Dec is John Herschel's *General Catalogue of Non-stellar Objects (GC)* of 1864, but this (equivalent) alternative has come more and more out of use since, so that virtually all current astronomical databases use Dec.

It remains to fix the zero point of the longitudinal coordinate, called Right Ascension (RA). For this, the intersection points of the equatorial plane with Earth's orbital plane, the *ecliptic*, are taken, more precisely the so-called *vernal equinox* or "First Point of Aries". During the year, as Earth moves around the Sun, the Sun *appears* to move through this point each year around March 21 when spring begins on the Northern hemisphere, and crosses the celestial equator from south to north (Southerners are asked to forgive a certain amount of "hemispherism" in the official nomenclature). The opposite point is called the "autumnal equinox", and the Sun passes it around September 23 when it returns to the Southern celestial hemisphere. As a longitudinal coordinate, RA can take values between 0 and 360 deg. However, this coordinate is more often given in time units hours (h), minutes (m), and seconds (s), where 24 hours correspond to 360 degrees (so that RA takes values between 0 and 24 h); the correspondence of units is as follows:

$$24 \text{ h} = 360 \text{ deg}$$

$$1 \text{ h} = 15 \text{ deg}, 1 \text{ m} = 15', 1 \text{ s} = 15''$$

$$1 \text{ deg} = 4 \text{ m}, 1' = 4 \text{ s}$$

So the vernal equinox, where the Sun appears to be when Northern spring begins around March 21, is at RA = 0 h = 0 deg, the summer solstice where the Sun is when Northern summer begins around June 21, is at RA = 6 h = 90 deg, the autumnal equinox is at RA = 12 h = 180 deg, and the winter solstice is at RA = 18 h = 270 deg. Thus RA is measured from west to east in the celestial sphere.

Because of small periodic and secular changes of the rotation axis of Earth, especially precession, the vernal equinox is not constant but varies slowly, so that the whole equatorial coordinate system is slowly changing with time. Therefore, it is necessary to give an epoch (a moment of time) for which the equatorial system is taken; currently, most sources use epoch 2000.0, the beginning of the year 2000 AD.

To go over from equatorial coordinates fixed to the stars to the horizon system, the concept of the *hour angle* (HA) is useful. In principle, this means introducing a new, *second* equatorial coordinate system which co-rotates with Earth. This system has again the celestial equator and poles as reference quantities, and declination as latitudinal coordinate, but a co-rotating longitudinal coordinate called hour angle. In this system, a star or other celestial object moves contrary to Earth's rotation along a circle of constant declination during the course of the day; This rotation leaves the celestial poles in the same invariant position for all time: They always stay on the *local meridian* of the observer (which goes through south and north point also), and the altitude of the north celestial pole is equal to the geographic latitude of the observer (thus negative for southerners, who cannot see it for this reason, but the south celestial pole instead). This meridian always coincides with in hour circle for this reason. Thus, as may be suggestive, the local meridian is taken as the hour circle for HA=0.

Celestial objects are at constant RA, but change their hour angle as time proceeds. If measured in units of hours, minutes and seconds, HA will change for the same amount as the elapsed time interval is, as measured in *star time* (ST), which is defined so that a siderial rotation of Earth takes 24 hours star time, which corresponds to 23 h 56 m 4.091 s standard (mean solar) time. This is actually the reason why RA and HA are measured in time units. The standard convention is that HA is measured from east to west so that it increases with time, and this is opposite to the convention for RA !

Star time is ST = 0 h by definition whenever the vernal equinox, RA = 0 h, crosses the local meridian, HA = 0. As time proceeds, RA stays constant, and both HA and ST grow by the amount of time elapsed, thus star time is always equal to the hour angle of the vernal equinox. Moreover, objects with "later" RA come into the meridian HA = 0, more precisely with RA which is later by the amount of elapsed star time, so that also star time is equal to the current Right Ascension of the local meridian.

More generally, for any object in the sky, the following relation between right ascension, hour angle, and star time always holds:

HA = ST - RA

(here given to determine the current HA from known RA and ST).

Transformation of Horizontal to Equatorial Coordinates, and Vice Versa. Measured observed coordinates in the horizontal system, azimuth A and altitude a, can be transformed to (co-rotating) equatorial ones, hour angle HA and declination Dec, for an observer at geographical latitude B, by the transformation formulae (mathematically, this is a rotation around the east-west axis by angle (90 deg - B)):

cos Dec * sin HA = cos a * sin A

sin Dec = sin B * sin a + cos B * cos a * cos A

$$\cos Dec * \cos HA = \cos B * \sin a + \sin B * \cos a * \cos A$$

The inverse transformation formulae from given HA, Dec to A, a read:

$$\cos a * \sin A = \cos Dec * \sin HA$$

$$\sin a = \sin B * \sin Dec + \cos B * \cos Dec * \cos HA$$

$$\cos a * \cos A = - \cos B * \sin Dec + \sin B * \cos Dec * \cos HA$$

For practical calculation in either case, evaluate e.g. the second formula first to obtain Dec or a, and then use the result in the first formula to get HA or A, respectively. (Get HA from or transform it to Right Ascension according to the relation given at the end of the last section if star time is known).

## Effects of Earth's Rotation

As mentioned above, Earth's (or another celestial body's) rotation has remarkable effects on the appearance of the sky: Stars and other celestial bodies appear to rotate around the celestial poles (as actually Earth rotates and carries the observer away below them), i.e. move along circles of constant declination in the co-rotating equatorial system.

By doing so, stars will cross the local meridian (defined e.g. by zero hour angle HA) twice a day; these events are called *transits* or *culminations*, i.e., the *upper* and the *lower* transit, or the *upper* and the *lower* culmination. These events also mark the maximal and minimal altitude a the objects can reach in the observer's sky, and may both take place above or below the horizon of the observer, depending on the declination Dec of the object and the geographic latitude $B$ of the observer.

The altitudes for *upper transits* are as follows:

$$a = 90 \text{ deg} - |B - Dec|$$

where the transit takes place *north* of the zenith if Dec > B and south otherwise. If |B - Dec| > 90 deg, the upper transit will take place at negative altitude, i.e. below the horizon, so that the object will never come above the horizon and thus never be visible; for the Northern hemisphere, this is true for all objects with,

$$Dec < B - 90 \text{ deg} (< 0),$$

and for the Southern hemisphere for,

$$Dec > B + 90 \text{ deg} (> 0).$$

The altitudes for the *lower transit* are given by,

$$a = (B + Dec) - 90 \text{ deg} \quad B > 0 \text{ (North)}$$

$$a = - (B + Dec) - 90 \text{ deg} \quad B < 0 \text{ (South)}$$

For an observer on the Northern hemisphere, stars with Dec > 90 deg - B (> 0), and for southern hemisphere observers, stars with Dec < - 90 deg - B (< 0) will have their lower transit at positive altitudes, i.e., above horizon, and will never set; such stars are called *circumpolar*.

All stars which are neither circumpolar nor never visible will have their upper transit above and their lower transit below horizon, and thus rise and set during a siderial day. Disregarding refraction effects, the hour angle of the rise and set of a celestial object, the *semidiurnal arc* H0, is given by,

$$\cos H0 = - \tan Dec * \tan B$$

while the azimuth of the rising and setting points, the evening and morning *elongation* A0 is,

$$\cos A0 = - \sin Dec / \cos B$$

where A0 > 90 dec if Dec and B have same sign (i.e., are on the same hemisphere). Rising and setting times differ from transit time by the amount of the diurnal arc H0, given in time units (hours), taken as hours of star time.

If Dec and B have same sign (i.e., are on the same hemisphere), one of the following situations occurs:

- If |Dec| < |B|, the object transits the prime vertical, A = +/- 90 deg; this occurs at altitude and hour angle given by,

$$\sin a = \sin Dec / \sin B$$

$$\cos HA = \tan Dec * \cot B$$

- If |Dec| > |B|, the object will stay within a certain region of azimuth around the visible celestial pole, where the extremal azimuth points are given by,

$$\sin a = \sin B / \sin Dec$$

$$\cos HA = \cot Dec * \tan B$$

## The Ecliptical Coordinate System

In the *ecliptical coordinate system*, the fundamental reference plane is chosen to be the *ecliptic*, i.e. the orbital plane of the Earth around the Sun. Earth's revolution around the Sun defines an orientation and thus the *North* and the *South Ecliptic Pole*.

The *ecliptic latitude* (be) is defined as the angle between a position and the ecliptic and takes values between -90 and +90 deg, while the *ecliptic longitude* (le) is again starting from the vernal equinox and runs from 0 to 360 deg in the same eastward sense as Right Ascension.

The obliquity, or inclination of Earth's equator against the ecliptic, amounts eps[ilon] = 23deg 26' 21.448" (2000.0) and changes very slightly with time, due to gravitational perturbations of Earth's motion. Knowing this quantity, the transformation formulae from equatorial to ecliptical coordinates are quite simply given (mathematically, by a rotation around the "X" axis pointing to the vernal equinox by angle eps):

cos be * cos le = cos Dec * cos RA

cos be * sin le = cos Dec * sin RA * cos eps + sin Dec * sin eps

sin be = - cos Dec * sin RA * sin eps + sin Dec * cos eps

and the reverse transformation:

cos Dec * cos RA = cos be * cos le

cos Dec * sin RA = cos be * sin le * cos eps - sin be * sin eps

sin Dec = cos be * sin le * sin eps + sin be * cos eps

Ecliptical coordinates are most frequently used for solar system calculations such as planetary and cometary orbits and appearances. For this purpose, two ecliptical systems are used: The heliocentric coordinate system with the Sun in its center, and the geocentric one with the Earth in its origin, which can be transferred into each other by a coordinate translation.

## Galactic Coordinates

This coordinate system is most useful for considerations of objects beyond the solar system, especially for considerations of objects of our Milky Way galaxy, and sometimes beyond.

Here, the galactic plane, or galactic equator, is used as reference plane. This is the great circle of the celestial sphere which best approximates the visible Milky Way. For historical reasons, the direction from us to the Galactic Center has been selected as zero point for *galactic longitude* l, and this was counted toward the direction of our Sun's rotational motion which is therefore at l = 90 deg. This sense of rotation, however, is opposite to the sense of rotation of our Galaxy, as can be easily checked ! Therefore, the *galactic north pole*, defined by the *galactic coordinate system*, coincides with the rotational south pole of our Galaxy, and vice versa.

*Galactic latitude* b is the angle between a position and the galactic equator and runs from -90 to +90 deg. Glalactic longitude runs of course from 0 to 360 deg.

The galactic north pole is at RA = 12:51.4, Dec = +27:07 (2000.0), the galactic center at RA = 17:45.6, Dec = -28:56 (2000.0). The inclination of the galactic equator to Earth's equator is thus 62.9 deg. The intersection, or node line of the two equators is at RA = 18:51.4, Dec = 0:00 (2000.0), and at l = 33 deg, b=0.

The transformation formulae for this frame get more complicated, as the transformation is consisted of, (1) a rotation around the celestial polar axis by 18:51.4 hours, so that the reference zero longitude matches the node, (2) a rotation around the node by 62.9 deg, followed by (3) a rotation around the galactic polar axis by 33 deg so that the zero longitude meridian matches the galactic center. This complicated transformation will not be given here formally.

Before 1959, the intersection line had been taken as zero galactic longitude, so that the old differed from the new latitude by 33.0 deg (the longitude of the node just discussed, but for the celestial equator of the epoch 1950.0):

l(old) = l(new) - 33.0 deg

For a transition time, the old coordinate had been assigned a superscript "I", the new longitude a superscript "II", which can be found in some literature.

For some considerations, besides the geo- or heliocentric galactic coordinates described above, *galactocentric galactic coordinates* are useful, which have the galactic center in their origin; these can be obtained from the helio/geocentric ones by a parallel translation.

## Astronomical Unit

The astronomical unit (symbol: au, ua, or AU) is a unit of length, roughly the distance from Earth to the Sun. However, that distance varies as Earth orbits the Sun, from a maximum (aphelion) to a minimum (perihelion) and back again once a year. Originally conceived as the average of Earth's aphelion and perihelion, since 2012 it has been defined as exactly $1.49597870700 \times 10^{11}$ metres, or about 150 million kilometres (93 million miles). The astronomical unit is used primarily for measuring distances within the Solar System or around other stars. It is also a fundamental component in the definition of another unit of astronomical length, the parsec.

A variety of unit symbols and abbreviations have been in use for the astronomical unit. In a 1976 resolution, the International Astronomical Union (IAU) had used the symbol $A$ to denote a length equal to the astronomical unit. In the astronomical literature, the symbol AU was (and remains) common. In 2006, the International Bureau of Weights and Measures (BIPM) had recommended ua as the symbol for the unit. In the non-normative Annex C to ISO 80000-3 (2006), the symbol of the astronomical unit is "ua". In

2012, the IAU, noting "that various symbols are presently in use for the astronomical unit", recommended the use of the symbol "au", as did the American Astronomical Society (AAS) in the manuscript preparation guidelines for its principal journals. In the 2014 revision and 2019 edition of the SI Brochure, the BIPM used the unit symbol "au".

## Development of Unit Definition

Earth's orbit around the Sun is an ellipse. The semi-major axis of this elliptic orbit is defined to be half of the straight line segment that joins the perihelion and aphelion. The centre of the Sun lies on this straight line segment, but not at its midpoint. Because ellipses are well-understood shapes, measuring the points of its extremes defined the exact shape mathematically, and made possible calculations for the entire orbit as well as predictions based on observation. In addition, it mapped out exactly the largest straight-line distance that Earth traverses over the course of a year, defining times and places for observing the largest parallax (apparent shifts of position) in nearby stars. Knowing Earth's shift and a star's shift enabled the star's distance to be calculated. But all measurements are subject to some degree of error or uncertainty, and the uncertainties in the length of the astronomical unit only increased uncertainties in the stellar distances. Improvements in precision have always been a key to improving astronomical understanding. Throughout the twentieth century, measurements became increasingly precise and sophisticated, and ever more dependent on accurate observation of the effects described by Einstein's theory of relativity and upon the mathematical tools it used.

Improving measurements were continually checked and cross-checked by means of improved understanding of the laws of celestial mechanics, which govern the motions of objects in space. The expected positions and distances of objects at an established time are calculated (in AU) from these laws, and assembled into a collection of data called an ephemeris. NASA's Jet Propulsion Laboratory HORIZONS System provides one of several ephemeris computation services.

In 1976, in order to establish a yet more precise measure for the astronomical unit, the IAU formally adopted a new definition. Although directly based on the then-best available observational measurements, the definition was recast in terms of the then-best mathematical derivations from celestial mechanics and planetary ephemerides. It stated that "the astronomical unit of length is that length ($A$) for which the Gaussian gravitational constant ($k$) takes the value 0.01720209895 when the units of measurement are the astronomical units of length, mass and time". Equivalently, by this definition, one AU is "the radius of an unperturbed circular Newtonian orbit about the sun of a particle having infinitesimal mass, moving with an angular frequency of 0.01720209895 radians per day"; or alternatively that length for which the heliocentric gravitational constant (the product $GM_\odot$) is equal to $(0.01720209895)^2$ AU$^3$/d$^2$, when the length is used to describe the positions of objects in the Solar System.

Subsequent explorations of the Solar System by space probes made it possible to obtain precise measurements of the relative positions of the inner planets and other objects by means of radar and telemetry. As with all radar measurements, these rely on measuring the time taken for photons to be reflected from an object. Because all photons move at the speed of light in vacuum, a fundamental constant of the universe, the distance of an object from the probe is calculated as the product of the speed of light and the measured time. However, for precision the calculations require adjustment for things such as the motions of the probe and object while the photons are transiting. In addition, the measurement of the time itself must be translated to a standard scale that accounts for relativistic time dilation. Comparison of the ephemeris positions with time measurements expressed in the TDB scale leads to a value for the speed of light in astronomical units per day (of 86400 s). By 2009, the IAU had updated its standard measures to reflect improvements, and calculated the speed of light at 173.1446326847(69) au/d (TDB).

In 1983, the International Committee for Weights and Measures (CIPM) modified the International System of Units (SI, or "modern" metric system) to make the metre defined as the distance travelled in a vacuum by light in 1/299792458 second. This replaced the previous definition, valid between 1960 and 1983, which was that the metre equalled a certain number of wavelengths of a certain emission line of krypton-86. (The reason for the change was an improved method of measuring the speed of light.) The speed of light could then be expressed exactly as $c_0$ = 299792458 m/s, a standard also adopted by the IERS numerical standards. From this definition and the 2009 IAU standard, the time for light to traverse an AU is found to be $\tau_A$ = 499.0047838061±0.00000001 s, which is slightly more than 8 minutes 19 seconds. By multiplication, the best IAU 2009 estimate was $A = c_0\tau_A$ = 149597870700±3 m, based on a comparison of JPL and IAA–RAS ephemerides.

In 2006, the BIPM reported a value of the astronomical unit as $1.49597870691(6) \times 10^{11}$ m. In the 2014 revision of the SI Brochure, the BIPM recognised the IAU's 2012 redefinition of the astronomical unit as 149597870700 m. or an increase of 9 meters.

This estimate was still derived from observation and measurements subject to error, and based on techniques that did not yet standardize all relativistic effects, and thus were not constant for all observers. In 2012, finding that the equalization of relativity alone would make the definition overly complex, the IAU simply used the 2009 estimate to redefine the astronomical unit as a conventional unit of length directly tied to the metre (exactly 149597870700 m). The new definition also recognizes as a consequence that the astronomical unit is now to play a role of reduced importance, limited in its use to that of a convenience in some applications.

| 1 astronomical unit | = 149597870700 metres (exactly) |
|---|---|
| | ≈ 92.955807 million miles |
| | ≈ 499.00478384 light-seconds |
| | ≈ 4.8481368 millionths ($4.8481368 \times 10^{-6}$) of a parsec |
| | ≈ 15.812507 millionths ($15.812507 \times 10^{-6}$) of a light-year |

This definition makes the speed of light, defined as exactly 299792458 m/s, equal to exactly 299792458 × 86400 ÷ 149597870700 or about 173.144632674240 AU/d, some 60 parts per trillion less than the 2009 estimate.

## Usage and Significance

With the definitions used before 2012, the astronomical unit was dependent on the heliocentric gravitational constant, that is the product of the gravitational constant $G$ and the solar mass $M_\odot$. Neither $G$ nor $M_\odot$ can be measured to high accuracy separately, but the value of their product is known very precisely from observing the relative positions of planets (Kepler's Third Law expressed in terms of Newtonian gravitation). Only the product is required to calculate planetary positions for an ephemeris, so ephemerides are calculated in astronomical units and not in SI units.

The calculation of ephemerides also requires a consideration of the effects of general relativity. In particular, time intervals measured on Earth's surface (terrestrial time, TT) are not constant when compared to the motions of the planets: the terrestrial second (TT) appears to be longer during the Northern Hemisphere winter and shorter during the Northern Hemisphere summer when compared to the "planetary second" (conventionally measured in barycentric dynamical time, TDB). This is because the distance between Earth and the Sun is not fixed (it varies between 0.9832898912 and 1.0167103335 AU) and, when Earth is closer to the Sun (perihelion), the Sun's gravitational field is stronger and Earth is moving faster along its orbital path. As the metre is defined in terms of the second and the speed of light is constant for all observers, the terrestrial metre appears to change in length compared to the "planetary metre" on a periodic basis.

The metre is defined to be a unit of proper length, but the SI definition does not specify the metric tensor to be used in determining it. Indeed, the International Committee for Weights and Measures (CIPM) notes that "its definition applies only within a spatial extent sufficiently small that the effects of the non-uniformity of the gravitational field can be ignored". As such, the metre is undefined for the purposes of measuring distances within the Solar System. The 1976 definition of the astronomical unit was incomplete because it did not specify the frame of reference in which time is to be measured, but proved practical for the calculation of ephemerides: a fuller definition that is consistent with general relativity was proposed, and "vigorous debate" ensued until August 2012 when the IAU adopted the current definition of 1 astronomical unit = 149597870700 metres.

The astronomical unit is typically used for stellar system scale distances, such as the size of a protostellar disk or the heliocentric distance of an asteroid, whereas other units are used for other distances in astronomy. The astronomical unit is too small to be convenient for interstellar distances, where the parsec and light-year are widely used. The parsec (parallax arcsecond) is defined in terms of the astronomical unit, being the distance of an object with a parallax of 1 arcsecond. The light-year is often used in popular works, but is not an approved non-SI unit and is rarely used by professional astronomers.

When simulating a numerical model of the Solar System, the astronomical unit provides an appropriate scale that minimizes (overflow, underflow and truncation) errors in floating point calculations.

## Developments

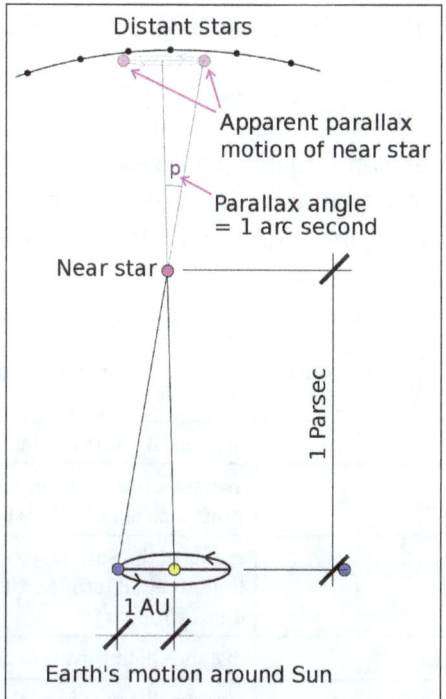

The astronomical unit is used as the baseline of the triangle to measure stellar parallaxes (distances in the image are not to scale).

The unit distance $A$ (the value of the astronomical unit in metres) can be expressed in terms of other astronomical constants:

$$A^3 = \frac{GM_\odot D^2}{k^2}$$

where $G$ is the Newtonian gravitational constant, $M_\odot$ is the solar mass, $k$ is the numerical value of Gaussian gravitational constant and $D$ is the time period of one day. The Sun is constantly losing mass by radiating away energy, so the orbits of the planets are steadily expanding outward from the Sun. This has led to calls to abandon the astronomical unit as a unit of measurement.

As the speed of light has an exact defined value in SI units and the Gaussian gravitational constant $k$ is fixed in the astronomical system of units, measuring the light time per unit distance is exactly equivalent to measuring the product $GM_\odot$ in SI units. Hence, it is possible to construct ephemerides entirely in SI units, which is increasingly becoming the norm.

A 2004 analysis of radiometric measurements in the inner Solar System suggested that the secular increase in the unit distance was much larger than can be accounted for by solar radiation, +15±4 metres per century.

The measurements of the secular variations of the astronomical unit are not confirmed by other authors and are quite controversial. Furthermore, since 2010, the astronomical unit has not been estimated by the planetary ephemerides.

## Examples

The following table contains some distances given in astronomical units. It includes some examples with distances that are normally not given in astronomical units, because they are either too short or far too long. Distances normally change over time. Examples are listed by increasing distance.

| Object | Length or distance (AU) | Range | Comment and reference point |
|---|---|---|---|
| Light-second | 0.002 | – | distance light travels in one second |
| Lunar distance | 0.0026 | – | average distance from Earth (which the Apollo missions took about 3 days to travel) |
| Solar radius | 0.005 | – | radius of the Sun (695500 km, 432450 mi, a hundred times the radius of Earth or ten times the average radius of Jupiter) |
| Light-minute | 0.12 | – | distance light travels in one minute |
| Mercury | 0.39 | – | average distance from the Sun |
| Venus | 0.72 | – | average distance from the Sun |
| Earth | 1.00 | – | average distance of Earth's orbit from the Sun (sunlight travels for 8 minutes and 19 seconds before reaching Earth) |
| Mars | 1.52 | – | average distance from the Sun |
| Light-hour | 7.2 | – | distance light travels in one hour |
| Kuiper belt | 30 | – | Inner edge begins at roughly 30 AU |
| Eris | 67.8 | – | average distance from the Sun |
| Light-day | 173 | – | distance light travels in one day |
| Light-year | 63241 | – | distance light travels in one Julian year (365.25 days) |
| Oort cloud | 75000 | ± 25000 | distance of the outer limit of Oort cloud from the Sun (estimated, corresponds to 1.2 light-years) |
| Parsec | 206265 | – | one parsec (The parsec is defined in terms of the astronomical unit, is used to measure distances beyond the scope of the Solar System and is about 3.26 light-years.) |
| Proxima Centauri | 268000 | ± 126 | distance to the nearest star to the Solar System |
| Galactic Centre | 1700000000 | – | distance from the Sun to the centre of the Milky Way |

## Astronomical Unit of Mass

### Solar Mass

The solar mass ($M_\odot$) is a standard unit of mass in astronomy, equal to approximately $2 \times 10^{30}$ kg. It is used to indicate the masses of other stars, as well as clusters, nebulae, and galaxies. It is equal to the mass of the Sun (denoted by the solar symbol $\odot$). This equates to about two nonillion (two quintillion in the long scale) kilograms:

$$M_\odot = (1.98847 \pm 0.00007) \times 10^{30} \text{ kg}$$

The above mass is about 332946 times the mass of Earth ($M_\oplus$), or 1047 times the mass of Jupiter ($M_J$).

Because Earth follows an elliptical orbit around the Sun, the solar mass can be computed from the equation for the orbital period of a small body orbiting a central mass. Based upon the length of the year, the distance from Earth to the Sun (an astronomical unit or AU), and the gravitational constant ($G$), the mass of the Sun is given by:

$$M_\odot = \frac{4\pi^2 \times (1\text{AU})^3}{G \times (1\text{yr})^2}$$

The value of $G$ is difficult to measure and is only known with limited accuracy in SI units. The value of $G$ times the mass of an object, called the standard gravitational parameter, is known for the Sun and several planets to much higher accuracy than $G$ alone. As a result, the solar mass is used as the standard mass in the astronomical system of units.

The value of the gravitational constant was first derived from measurements that were made by Henry Cavendish in 1798 with a torsion balance. The value he obtained differs by only 1% from the modern value. The diurnal parallax of the Sun was accurately measured during the transits of Venus in 1761 and 1769, yielding a value of 9″ (9 arcseconds, compared to the present 1976 value of 8.794148″). From the value of the diurnal parallax, one can determine the distance to the Sun from the geometry of Earth.

The first person to estimate the mass of the Sun was Isaac Newton. In his work *Principia* (1687), he estimated that the ratio of the mass of Earth to the Sun was about 1/28 700. Later he determined that his value was based upon a faulty value for the solar parallax, which he had used to estimate the distance to the Sun (1 AU). He corrected his estimated ratio to 1/169 282 in the third edition of the *Principia*. The current value for the solar parallax is smaller still, yielding an estimated mass ratio of 1/332 946.

As a unit of measurement, the solar mass came into use before the AU and the gravitational constant were precisely measured. This is because the relative mass of another planet in the Solar System or the combined mass of two binary stars can be calculated in units of Solar mass directly from the orbital radius and orbital period of the planet or stars using Kepler's third law, provided that orbital radius is measured in astronomical units and orbital period is measured in years.

The mass of the Sun has been decreasing since the time it formed. This occurs through two processes in nearly equal amounts. First, in the Sun's core, hydrogen is converted into helium through nuclear fusion, in particular the p–p chain, and this reaction converts some mass into energy in the form of gamma ray photons. Most of this energy eventually radiates away from the Sun. Second, high-energy protons and electrons in the atmosphere of the Sun are ejected directly into outer space as the solar wind and coronal mass ejections.

The original mass of the Sun at the time it reached the main sequence remains uncertain. The early Sun had much higher mass-loss rates than at present, and it may have lost anywhere from 1–7% of its natal mass over the course of its main-sequence lifetime. The Sun gains a very small amount of mass through the impact of asteroids and comets. However, as the Sun already contains 99.86% of the Solar System's total mass, these impacts cannot offset the mass lost by radiation and ejection.

## Related Units

One solar mass, $M_\odot$, can be converted to related units:

- 27068510 $M_L$ (Lunar mass).

- 332946 $M_\oplus$ (Earth mass).

- 1047.35 $M_J$ (Jupiter mass).

- 1988.55 yottatonnes.

It is also frequently useful in general relativity to express mass in units of length or time:

- $M_\odot G / c^2 \approx 1.48$ km (half the Schwarzschild radius of the Sun).

- $M_\odot G / c^3 \approx 4.93$ μs.

The solar mass parameter $(G \cdot M_\odot)$, as listed by the IAU Division I Working Group, has the following estimates:

- $1.32712442099 \times 10^{20}$ m³s⁻² (TCG-compatible).

- $1.32712440041 \times 10^{20}$ m³s⁻² (TDB-compatible).

## Jupiter Mass

Jupiter mass, also called Jovian mass, is the unit of mass equal to the total mass of the planet Jupiter. This value may refer to the mass of the planet alone, or the mass of the entire Jovian system to include the moons of Jupiter. Jupiter is by far the most massive planet in the Solar System. It is approximately 2.5 times as massive as all of the other planets in the Solar System combined.

Jupiter mass is a common unit of mass in astronomy that is used to indicate the masses of other similarly-sized objects, including the outer planets and extrasolar planets. It may also be used to describe the masses of brown dwarfs, as this unit provides a convenient scale for comparison.

## Current best Estimates

The current best known value for the mass of Jupiter can be expressed as 1898130 yottagrams:

$$M_J = (1.89813 \pm 0.00019) \times 10^{27} \text{ kg,}$$

which is about $\frac{1}{1000}$ as massive as the sun (is about 0.1% $M_\odot$):

$$M_J = \frac{1}{1047.348644 \pm 0.000017} M_{Sun} \approx (9.547919 \pm 0.000002) \times 10^{-4} M_{Sun}.$$

Jupiter is 318 times as massive as Earth:

$$M_J = 317.82838 \, M_\oplus$$

## Context and Implications

Jupiter's mass is 2.5 times that of all the other planets in the Solar System combined—this is so massive that its barycenter with the Sun lies beyond the Sun's surface at 1.068 solar radii from the Sun's center.

Because the mass of Jupiter is so large compared to the other objects in the solar system, the effects of its gravity must be included when calculating satellite trajectories and the precise orbits of other bodies in the solar system, including Earth's moon and even Pluto.

Theoretical models indicate that if Jupiter had much more mass than it does at present, its atmosphere would collapse, and the planet would shrink. For small changes in mass, the radius would not change appreciably, but above about 500 $M_\oplus$ (1.6 Jupiter masses) the interior would become so much more compressed under the increased pressure that its volume would *decrease* despite the increasing amount of matter. As a result, Jupiter is thought to have about as large a diameter as a planet of its composition and evolutionary history can achieve. The process of further shrinkage with increasing mass would continue until appreciable stellar ignition was achieved, as in high-mass brown dwarfs having around 50 Jupiter masses. Jupiter would need to be about 75 times as massive to fuse hydrogen and become a star.

## Gravitational Constant

The mass of Jupiter is derived from the measured value called the Jovian mass parameter, which is denoted with $GM_J$. The mass of Jupiter is calculated by dividing $GM_J$ by the constant $G$. For celestial bodies such as Jupiter, Earth and the Sun, the value of the $GM$ product is known to many orders of magnitude more precisely than either factor independently. The limited precision available for $G$ limits the uncertainty of the derived mass. For this reason, astronomers often prefer to refer to the gravitational parameter, rather than the explicit mass. The $GM$ products are used when computing the ratio of Jupiter mass relative to other objects.

In 2015, the International Astronomical Union defined the *nominal Jovian mass parameter* to remain constant regardless of subsequent improvements in measurement precision of $M_J$. This constant is defined as exactly,

$$(\mathcal{GM})_J^N = 1.2668653 \times 10^{17} \text{ m}^3 / \text{s}^2$$

If the explicit mass of Jupiter is needed in SI units, it can be calculated in terms of the gravitational constant, $G$ by dividing $GM$ by $G$.

## Mass Composition

The majority of Jupiter's mass is hydrogen and helium. These two elements make up more than 87% of the total mass of Jupiter. The total mass of heavy elements other than hydrogen and helium in the planet is between 11 and 45 $M_\oplus$. The bulk of the hydrogen on Jupiter is solid hydrogen. Evidence suggests that Jupiter contains a central dense core. If so, the mass of the core is predicted to be no larger than about 12 $M_\oplus$. The exact mass of the core is uncertain due to the relatively poor knowledge of the behavior of solid hydrogen at very high pressures.

## Relative Mass

| Masses of noteworthy astronomical objects relative to the mass of Jupiter | | |
|---|---|---|
| Object | $M_J / M_{object}$ | $M_{object} / M_J$ |
| Sun | 954.7919(15)×10⁻⁶ | 1047.348644±0.000017 |
| Earth | 317.82838 | 0.0031463520 |
| Jupiter | 1 | 1 |
| Saturn | 3.3397683 | 0.29942197 |
| Uranus | 21.867552 | 0.045729856 |
| Neptune | 18.53467 | 0.05395295 |
| Gliese 229B | | 21–52.4 |
| 51 Pegasi b | | 0.472±0.039 |

## Earth Mass

Earth mass ($M_E$ or $M_\oplus$, where $\oplus$ is the standard astronomical symbol for planet Earth) is the unit of mass equal to that of Earth. The current best estimate for Earth mass is $M_\oplus$ = 5.9722×10²⁴ kg, with a standard uncertainty of 6×10²⁰ kg (relative uncertainty 10⁻⁴). It is equivalent to an average density of 5515 kg·m⁻³.

The Earth mass is a standard unit of mass in astronomy that is used to indicate the masses of other planets, including rocky terrestrial planets and exoplanets. One Solar mass is close to 333,000 Earth masses. The Earth mass excludes the mass of the Moon. The mass of the Moon is about 1.2% of that of the Earth, so that the mass of the Earth+Moon system is close to 6.0456×10²⁴ kg.

Most of the mass is accounted for by iron and oxygen (c. 32% each), magnesium and silicon (c. 15% each), calcium, aluminium and nickel (c. 1.5% each).

Precise measurement of the Earth mass is difficult, as it is equivalent to measuring the gravitational constant, which is the fundamental physical constant known with least accuracy, due to the relative weakness of the gravitational force. The mass of the Earth was first measured with any accuracy (within about 20% of the correct value) in the Schiehallion experiment in the 1770s, and within 1% of the modern value in the Cavendish experiment of 1798.

## Unit of Mass in Astronomy

The mass of Earth is estimated to be:

$$M_\oplus = (5.9722 \pm 0.0006) \times 10^{24} \text{ kg},$$

which can be expressed in terms of solar mass as:

$$M_\oplus = \frac{1}{332\,946.0487 \pm 0.0007} M_\odot \approx 3.003 \times 10^{-6} \, M_\odot.$$

The ratio of Earth mass to lunar mass has been measured to great accuracy. The current best estimate is:

$$M_\oplus / M_L = 81.3005678 \pm 0.0000027$$

| Masses of noteworthy astronomical objects relative to the mass of Earth | |
|---|---|
| Object | Earth mass $M_\oplus$ |
| Moon | 0.0123000371(4) |
| Sun | 332946.0487±0.0007 |
| Mercury | 0.0553 |
| Venus | 0.815 |

| Earth | 1 |
|---|---|
| Mars | 0.107 |
| Jupiter | 317.8 |
| Saturn | 95.2 |
| Uranus | 14.5 |
| Neptune | 17.1 |
| Gliese 667 Cc | 3.8 |
| Kepler-442b | 1.0 – 8.2 |

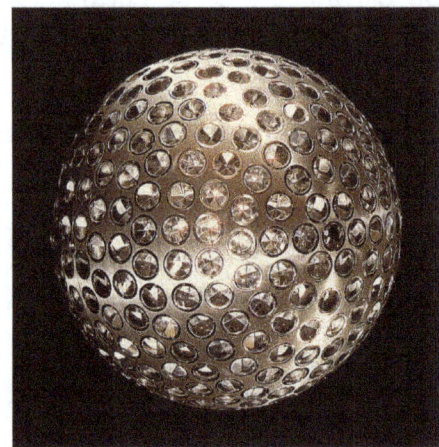

The LAGEOS satellite was used to precisely measure Earth's gravity, and therefore, its mass.

The $GM_\oplus$ product for the Earth is called the geocentric gravitational constant and equals $(398600441.8\pm0.8)\times10^6$ m$^3$ s$^{-2}$. It is determined using laser ranging data from Earth-orbiting satellites, such as LAGEOS-1. The $GM_\oplus$ product can also be calculated by observing the motion of the Moon or the period of a pendulum at various elevations. These methods are less precise than observations of artificial satellites.

The relative uncertainty of the geocentric gravitational constant is just $2\times10^{-9}$, i.e. 50000 times smaller than the relative uncertainty for $M_\oplus$ itself. $M_\oplus$ can be found out only by dividing the $GM_\oplus$ product by $G$, and $G$ is known only to a relative uncertainty of $4.6\times10^{-5}$ (2014 NIST recommended value), so $M_\oplus$ will have the same uncertainty at best. For this reason and others, astronomers prefer to use the un-reduced $GM_\oplus$ product, or mass ratios (masses expressed in units of Earth mass or Solar mass) rather than mass in kilograms when referencing and comparing planetary objects.

## Composition

Earth's density varies considerably, between less than 2700 kg·m$^{-3}$ in the upper crust to as much as 13000 kg·m$^{-3}$ in the inner core. The Earth's core accounts for 15% of Earth's volume but more than 30% of the mass, the mantle for 84% of the volume and close to 70% of the mass, while the crust accounts for less than 1% of the mass. About 90% of the mass of the Earth is composed of the iron–nickel alloy (95% iron) in the core

(30%), and the silicon dioxides (c. 33%) and magnesium oxide (c. 27%) in the mantle and crust. Minor contributions are from iron(II) oxide (5%), aluminium oxide (3%) and calcium oxide (2%), besides numerous trace elements (in elementary terms: iron and oxygen c. 32% each, magnesium and silicon c. 15% each, calcium, aluminium and nickel c. 1.5% each). Carbon accounts for 0.03%, water for 0.02%, and the atmosphere for about one part per million.

## Measurement

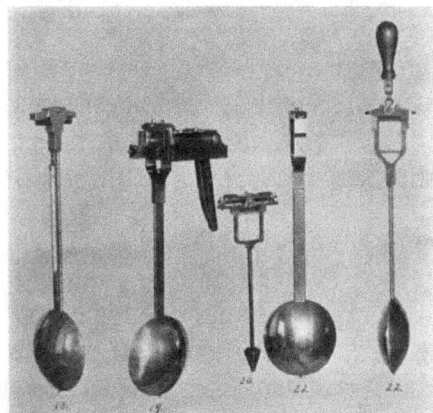

Pendulums used in Mendenhall gravimeter apparatus, from 1897 scientific journal. The portable gravimeter developed in 1890 by Thomas C. Mendenhall provided the most accurate relative measurements of the local gravitational field of the Earth.

The mass of Earth is measured indirectly by determining other quantities such as Earth's density, gravity, or gravitational constant. The first measurement in the 1770s Schiehallion experiment resulted in a value about 20% too low. The Cavendish experiment of 1798 found the correct value within 1%. Uncertainty was reduced to about 0.2% by the 1890s, to 0.1% by 1930.

The figure of the Earth has been known to better than four significant digits since the 1960s (WGS66), so that since that time, the uncertainty of the Earth mass is determined essentially by the uncertainty in measuring the gravitational constant. Relative uncertainty was cited at 0.06% in the 1970s, and at 0.01% ($10^{-4}$) by the 2000s. The current relative uncertainty of $10^{-4}$ amounts to $6\times10^{20}$ kg in absolute terms, of the order of the mass of a minor planet (70% of the mass of Ceres).

## Early Estimates

Before the direct measurement of the gravitational constant, estimates of the Earth mass were limited to estimating Earth's mean density from observation of the crust and estimates on Earth's volume. Estimates on the volume of the Earth in the 17th century were based on a circumference estimate of 60 miles (97 km) to the degree of latitude, corresponding to a radius of 5,500 km (86% of the Earth's actual radius of about 6,371 km), resulting in an estimated volume of about one third smaller than the correct value.

The average density of the Earth was not accurately known. Earth was assumed to consist either mostly of water (Neptunism) or mostly of igneous rock (Plutonism), both suggesting average densities far too low, consistent with a total mass of the order of $10^{24}$ kg. Isaac Newton estimated, without access to reliable measurement, that the density of Earth would be five or six times as great as the density of water, which is surprisingly accurate (the modern value is 5.515). Newton under-estimated the Earth's volume by about 30%, so that his estimate would be roughly equivalent to $(4.2\pm0.5)\times10^{24}$ kg.

In the 18th century, knowledge of Newton's law of gravitation permitted indirect estimates on the mean density of the Earth, via estimates of (what in modern terminology is known as) the gravitational constant. Early estimates on the mean density of the Earth were made by observing the slight deflection of a pendulum near a mountain, as in the Schiehallion experiment. Newton considered the experiment in *Principia*, but pessimistically concluded that the effect would be too small to be measurable.

An expedition from 1737 to 1740 by Pierre Bouguer and Charles Marie de La Condamine attempted to determine the density of Earth by measuring the period of a pendulum (and therefore the strength of gravity) as a function of elevation. The experiments were carried out in Ecuador and Peru, on Pichincha Volcano and mount Chimborazo. Bouguer wrote in a 1749 paper that they had been able to detect a deflection of 8 seconds of arc, the accuracy was not enough for a definite estimate on the mean density of the Earth, but Bouguer stated that it was at least sufficient to prove that the Earth was not hollow.

## Schiehallion Experiment

That a further attempt should be made on the experiment was proposed to the Royal Society in 1772 by Nevil Maskelyne, Astronomer Royal. He suggested that the experiment would "do honour to the nation where it was made" and proposed Whernside in Yorkshire, or the Blencathra-Skiddaw massif in Cumberland as suitable targets. The Royal Society formed the Committee of Attraction to consider the matter, appointing Maskelyne, Joseph Banks and Benjamin Franklin amongst its members. The Committee despatched the astronomer and surveyor Charles Mason to find a suitable mountain.

After a lengthy search over the summer of 1773, Mason reported that the best candidate was Schiehallion, a peak in the central Scottish Highlands. The mountain stood in isolation from any nearby hills, which would reduce their gravitational influence, and its symmetrical east–west ridge would simplify the calculations. Its steep northern and southern slopes would allow the experiment to be sited close to its centre of mass, maximising the deflection effect. Nevil Maskelyne, Charles Hutton and Reuben Burrow performed the experiment, completed by 1776. Hutton (1778) reported that the mean density of the Earth was estimated at that of Schiehallion mountain. This corresponds to a mean density about $4\frac{1}{2}$ higher than that of water (i.e., about 4.5 g/cm³), about 20%

below the modern value, but still significantly larger than the mean density of normal rock, suggesting for the first time that the interior of the Earth might be substantially composed of metal. Hutton estimated this metallic portion to occupy some $^{20}/_{31}$ (or 65%) of the diameter of the Earth (modern value 55%). With a value for the mean density of the Earth, Hutton was able to set some values to Jérôme Lalande's planetary tables, which had previously only been able to express the densities of the major Solar System objects in relative terms.

## Cavendish Experiment

The Henry Cavendish (1798) was the first to attempt to measure the gravitational attraction between two bodies directly in the laboratory. Earth's mass could be then found by combining two equations; Newton's second law, and Newton's law of universal gravitation.

In modern notation, the mass of the Earth is derived from the gravitational constant and the mean Earth radius by,

$$M_\oplus = \frac{GM_\oplus}{G} = \frac{gR_\oplus^2}{G}$$

Where "little g":

$$g = G\frac{M_\oplus}{R_\oplus^2}.$$

Cavendish found a mean density of 5.45 g/cm³, about 1% below the modern value.

## 19th Century

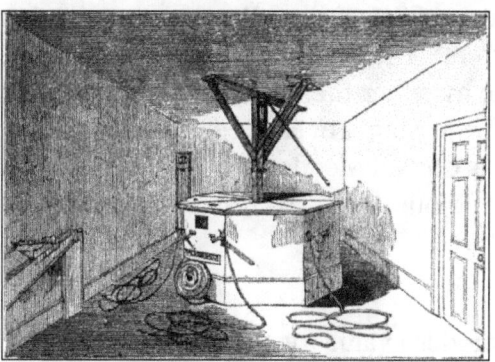

Experimental setup by Francis Baily and Henry Foster to determine the density of Earth using the Cavendish method.

While the mass of the Earth is implied by stating the Earth's radius and density, it was not usual to state the absolute mass explicitly prior to the introduction of scientific notation using powers of 10 in the later 19th century, because the absolute numbers would have been too awkward. Ritchie (1850) gives the mass of the Earth's atmosphere as "11,456,688,186,392,473,000 lbs." ($1.1\times10^{19}$ lb = $5.0\times10^{18}$ kg, modern value is $5.15\times10^{18}$ kg) and states that "compared with the weight of the globe this mighty sum dwindles to insignificance".

Absolute figures for the mass of the Earth are cited only beginning in the second half of the 19th century, mostly in popular rather than expert literature. An early such figure was given as "14 quadrillion pounds" (*14 Quadrillionen Pfund*) [$6.5\times10^{24}$ kg] in Masius (1859). Beckett (1871) cites the "weight of the earth" as "5842 quintillion tons" [$5.936\times10^{24}$ kg]. The "mass of the earth in gravitational measure" is stated as "$9.81996\times6370980^2$" with a "logarithm of earth's mass" given as "14.600522" [$3.98586\times10^{14}$]. This is the gravitational parameter in $m^3 \cdot s^{-2}$ (modern value $3.98600\times10^{14}$) and not the absolute mass.

Experiments involving pendulums continued to be performed in the first half of the 19th century. By the second half of the century, these were outperformed by repetitions of the Cavendish experiment, and the modern value of $G$ (and hence, of the Earth mass) is still derived from high-precision repetitions of the Cavendish experiment.

In 1821, Francesco Carlini determined a density value of $\rho$ = 4.39 g/cm³ through measurements made with pendulums in the Milan area. This value was refined in 1827 by Edward Sabine to 4.77 g/cm³, and then in 1841 by Carlo Ignazio Giulio to 4.95 g/cm³. On the other hand, George Biddell Airy sought to determine $\rho$ by measuring the difference in the period of a pendulum between the surface and the bottom of a mine. The first tests took place in Cornwall between 1826 and 1828. The experiment was a failure due to a fire and a flood. Finally, in 1854, Airy got the value 6.6 g/cm³ by measurements in a coal mine in Harton, Sunderland. Airy's method assumed that the Earth had a spherical stratification. Later, in 1883, the experiments conducted by Robert von Sterneck (1839 to 1910) at different depths in mines of Saxony and Bohemia provided the average density values $\rho$ between 5.0 and 6.3 g/cm³. This led to the concept of isostasy, which limits the ability to accurately measure $\rho$, by either the deviation from vertical of a plumb line or using pendulums. Despite the little chance of an accurate estimate of the average density of the Earth in this way, Thomas Corwin Mendenhall in 1880 realized a gravimetry experiment in Tokyo and at the top of Mount Fuji. The result was $\rho$ = 5.77 g/cm³.

## Modern Value

The uncertainty in the modern value for the Earth's mass has been entirely due to the uncertainty in the gravitational constant $G$ since at least the 1960s. $G$ is notoriously difficult to measure, and some high-precision measurements during the 1980s to 2010s have yielded mutually exclusive results. Sagitov (1969) based on the measurement of $G$ by Heyl and Chrzanowski (1942) cited a value of $M_\oplus$ = $5.973(3)\times10^{24}$ kg (relative uncertainty $5\times10^{-4}$).

Accuracy has improved only slightly since then. Most modern measurements are repetitions of the Cavendish experiment, with results (within standard uncertainty) ranging between 6.672 and 6.676 $\times 10^{-11}$ m$^3$ kg$^{-1}$s$^{-2}$ (relative uncertainty $3 \times 10^{-4}$) in results reported since the 1980s, although the 2014 NIST recommended value is close to $6.674 \times 10^{-11}$ m$^3$ kg$^{-1}$s$^{-2}$ with a relative uncertainty below $10^{-4}$. The *Astronomical Almanach Online* as of 2016 recommends a standard uncertainty of $1 \times 10^{-4}$ for Earth mass, $M_{\oplus}$ 5.9722(6)$\times 10^{24}$ kg.

## Variation

Earth's mass is variable, subject to both gain and loss due to the accretion of micro-meteorites and cosmic dust and the loss of hydrogen and helium gas, respectively. The combined effect is a net loss of material, estimated at $5.5 \times 10^7$ kg ($5.4 \times 10^4$ long tons) per year. This amount is $10^{-17}$ of the total earth mass. The $5.5 \times 10^7$ kg annual net loss is essentially due to 100,000 tons lost due to atmospheric escape, and an average of 45,000 tons gained from in-falling dust and meteorites. This is well within the mass uncertainty of 0.01% ($6 \times 10^{20}$ kg), so the estimated value of Earth's mass is unaffected by this factor.

Mass loss is due to atmospheric escape of gases. About 95,000 tons of hydrogen per year (3 kg/s) and 1,600 tons of helium per year are lost through atmospheric escape. The main factor in mass gain is in-falling material, cosmic dust, meteors, etc. are the most significant contributors to Earth's increase in mass. The sum of material is estimated to be 37000 to 78000 tons annually, although this can vary significantly; to take an extreme example, the Chicxulub impactor, with a midpoint mass estimate of $2.3 \times 10^{17}$ kg, added 900 million times that annual dustfall amount to the Earth's mass in a single event.

Additional changes in mass are due to the mass–energy equivalence principal, although these changes are relatively negligible. Mass loss due to the combination of nuclear fission and natural radioactive decay is estimated to amount to 16 tons per year, though these do not on their own change the total mass-energy of the earth.

An additional loss due to spacecraft on escape trajectories has been estimated at 65 tons per year since the mid-20th century. Earth lost about 3473 tons in the initial 53 years of the space age, but the trend is currently decreasing.

## Big Bang Theory

The big bang theory is the dominant theory of the origin of the universe. In essence, this theory states that the universe began from an initial point or singularity, which has expanded over billions of years to form the universe as we now know it.

## Early Expanding Universe Findings

In 1922, a Russian cosmologist and mathematician named Alexander Friedman found that solutions to Albert Einstein's general relativity field equations resulted in an expanding universe. As a believer in a static, eternal universe, Einstein added a cosmological constant to his equations, "correcting" for this "error" and thus eliminating the expansion. He would later call this the biggest blunder of his life.

Actually, there was already observational evidence in support of an expanding universe. In 1912, American astronomer Vesto Slipher observed a spiral galaxy—considered a "spiral nebula" at the time, since astronomers didn't yet know that there were galaxies beyond the Milky Way—and recorded its redshift, the shift of a light source shift toward the red end of the light spectrum. He observed that all such nebula were traveling away from the Earth. These results were quite controversial at the time, and their full implications were not considered at the time.

In 1924, astronomer Edwin Hubble was able to measure the distance to these "nebula" and discovered that they were so far away that they were not actually part of the Milky Way. He had discovered that the Milky Way was only one of many galaxies and that these "nebulae" were actually galaxies in their own right.

## Birth of the Big Bang

In 1927, Roman Catholic priest and physicist Georges Lemaitre independently calculated the Friedman solution and again suggested that the universe must be expanding. This theory was supported by Hubble when, in 1929, he found that there was a correlation between the distance of the galaxies and the amount of redshift in that galaxy's light. The distant galaxies were moving away faster, which was exactly what was predicted by Lemaitre's solutions.

In 1931, Lemaitre went further with his predictions, extrapolating backward in time find that the matter of the universe would reach an infinite density and temperature at a finite time in the past. This meant the universe must have begun in an incredibly small, dense point of matter, called a "primeval atom."

The fact that Lemaitre was a Roman Catholic priest concerned some, as he was putting forth a theory that presented a definite moment of "creation" to the universe. In the 1920s and 1930s, most physicists—like Einstein—were inclined to believe that the universe had always existed. In essence, the big-bang theory was seen as too religious by many people.

## Big Bang vs. Steady State

While several theories were presented for a time, it was really only Fred Hoyle's steady-state theory that provided any real competition for Lemaitre's theory. It was, ironically,

Hoyle who coined the phrase "Big Bang" during a 1950s radio broadcast, intending it as a derisive term for Lemaitre's theory.

The steady-state theory predicted that new matter was created such that the density and temperature of the universe remained constant over time, even while the universe was expanding. Hoyle also predicted that denser elements were formed from hydrogen and helium through the process of stellar nucleosynthesis, which, unlike the steady-state theory, has proved to be accurate.

George Gamow—one of Friedman's pupils—was the major advocate of the big-bang theory. Together with colleagues Ralph Alpher and Robert Herman, he predicted the cosmic microwave background (CMB) radiation, which is radiation that should exist throughout the universe as a remnant of the Big Bang. As atoms began to form during the recombination era, they allowed microwave radiation (a form of light) to travel through the universe, and Gamow predicted that this microwave radiation would still be observable today.

The debate continued until 1965 when Arno Penzias and Robert Woodrow Wilson stumbled upon the CMB while working for Bell Telephone Laboratories. Their Dicke radiometer, used for radio astronomy and satellite communications, picked up a 3.5 K temperature (a close match to Alpher and Herman's prediction of 5 K).

Throughout the late 1960s and early 1970s, some proponents of steady-state physics attempted to explain this finding while still denying the big-bang theory, but by the end of the decade, it was clear that the CMB radiation had no other plausible explanation. Penzias and Wilson received the 1978 Nobel Prize in physics for this discovery.

## Cosmic Inflation

Certain concerns, however, remained regarding the big-bang theory. One of these was the problem of homogeneity. Scientists asked: Why does the universe look identical, in terms of energy, regardless of which direction one looks? The big-bang theory does not give the early universe time to reach thermal equilibrium, so there should be differences in energy throughout the universe.

In 1980, American physicist Alan Guth formally proposed inflation theory to resolve this and other problems. This theory says that in the early moments following the Big Bang, there was an extremely rapid expansion of the nascent universe driven by "negative-pressure vacuum energy" (which *may* be in some way related to current theories of dark energy). Alternatively, inflation theories, similar in concept but with slightly different details have been put forward by others in the years since.

The Wilkinson Microwave Anisotropy Probe (WMAP) program by NASA, which began in 2001, has provided evidence that strongly supports an inflation period in the early universe. This evidence is especially strong in the three-year data released in 2006, though there are still some minor inconsistencies with theory. The 2006 Nobel Prize in Physics was awarded to John C. Mather and George Smoot, two key workers on the WMAP project.

## Existing Controversies

While the Big Bang theory is accepted by the vast majority of physicists, there are still some minor questions concerning it. Most importantly, however, are the questions which the theory cannot even attempt to answer:

- What existed before the Big Bang?

- What caused the Big Bang?

- Is our universe the only one?

The answers to these questions may well exist beyond the realm of physics, but they're fascinating nonetheless, and answers such as the multiverse hypothesis provide an intriguing area of speculation for scientists and non-scientists alike.

## Other Names for the Big Bang

When Lemaitre originally proposed his observation about the early universe, he called this early state of the universe the primeval atom. Years later, George Gamow would apply the name ylem for it. It has also been called the primordial atom or even the cosmic egg.

## Cosmic Microwave Background

The cosmic microwave background (CMB, CMBR), in Big Bang cosmology, is electromagnetic radiation as a remnant from an early stage of the universe, also known as "relic radiation". The CMB is faint cosmic background radiation filling all space. It is an important source of data on the early universe because it is the oldest electromagnetic radiation in the universe, dating to the epoch of recombination. With a traditional optical telescope, the space between stars and galaxies (the *background*) is completely dark. However, a sufficiently sensitive radio telescope shows a faint background noise, or glow, almost isotropic, that is not associated with any star, galaxy, or other object. This glow is strongest in the microwave region of the radio spectrum. The accidental discovery of the CMB in 1964 by American radio astronomers Arno Penzias and Robert Wilson was the culmination of work initiated in the 1940s, and earned the discoverers the 1978 Nobel Prize in Physics.

CMB is landmark evidence of the Big Bang origin of the universe. When the universe was young, before the formation of stars and planets, it was denser, much hotter, and filled with a uniform glow from a white-hot fog of hydrogen plasma. As the universe expanded, both the plasma and the radiation filling it grew cooler. When the universe cooled enough, protons and electrons combined to form neutral hydrogen atoms. Unlike the uncombined protons and electrons, these newly conceived atoms could not scatter the thermal radiation by Thomson scattering, and so the universe became transparent instead of being an opaque fog. Cosmologists refer to the time period when neutral atoms first formed as the *recombination epoch*, and the event shortly afterwards when photons started to travel freely through space rather than constantly being scattered by electrons and protons in plasma is referred to as photon decoupling. The photons that existed at the time of photon decoupling have been propagating ever since, though growing fainter and less energetic, since the expansion of space causes their wavelength to increase over time (and wavelength is inversely proportional to energy according to Planck's relation). This is the source of the alternative term *relic radiation*. The *surface of last scattering* refers to the set of points in space at the right distance from us so that we are now receiving photons originally emitted from those points at the time of photon decoupling.

Tiny residual variations in the glow show a very specific pattern, as would be expected of a fairly uniformly distributed hot gas that has expanded to the current size of the universe. In particular, the spectral radiance contains small anisotropies, or irregularities, which vary with the size of the region examined. They have been measured in detail, and match what would be expected if small thermal variations, generated by quantum fluctuations of matter in a very tiny space, had expanded to the size of the observable universe we see today. Although many different processes might produce the general form of a black body spectrum, no model other than the Big Bang has yet explained the fluctuations. As a result, most cosmologists consider the Big Bang model of the universe to be the best explanation for the CMB.

## Importance of Precise Measurement

Precise measurements of the CMB are critical to cosmology, since any proposed model of the universe must explain this radiation. The CMB has a thermal black body spectrum at a temperature of $2.72548 \pm 0.00057$ K. The spectral radiance $dE_\nu/d\nu$ peaks at 160.23 GHz, in the microwave range of frequencies, corresponding to a photon energy of about $6.626 \times 10^{-4}$ eV. Alternatively, if spectral radiance is defined as $dE_\lambda/d\lambda$, then the peak wavelength is 1.063 mm (282 GHz, $1.168 \times 10^{-3}$ eV photons). The glow is very nearly uniform in all directions, but the tiny residual variations show a very specific pattern, the same as that expected of a fairly uniformly distributed hot gas that has expanded to the current size of the universe. In particular, the spectral radiance at different angles of observation in the sky contains small anisotropies, or irregularities, which vary with the size of the region examined. They have been measured in detail, and match what would be expected if small thermal variations, generated by quantum fluctuations of matter in a very tiny space, had expanded to the size of the observable universe we see today. This is a very active field of study, with scientists seeking both better data (for example, the Planck spacecraft) and better interpretations of the initial conditions of expansion. Although many different processes might produce the general form of a black body spectrum, no model other than the Big Bang has yet explained the fluctuations. As a result, most cosmologists consider the Big Bang model of the universe to be the best explanation for the CMB.

The high degree of uniformity throughout the observable universe and its faint but measured anisotropy lend strong support for the Big Bang model in general and the ΛCDM ("Lambda Cold Dark Matter") model in particular. Moreover, the fluctuations are coherent on angular scales that are larger than the apparent cosmological horizon at recombination. Either such coherence is acausally fine-tuned, or cosmic inflation occurred.

## Features

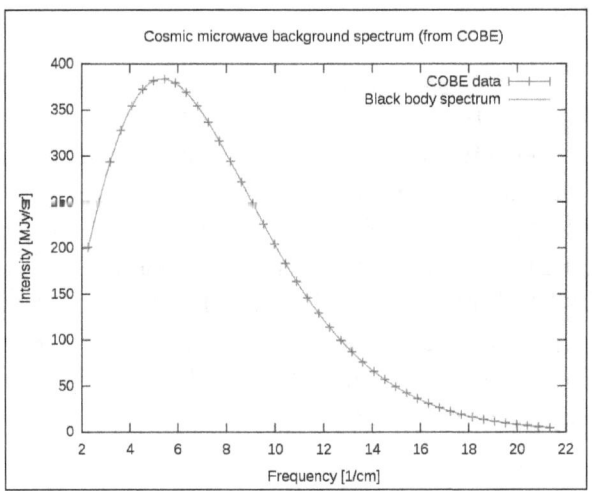

Graph of cosmic microwave background spectrum measured by the FIRAS instrument on the COBE, the most precisely measured black body spectrum in nature. The error bars are too small to be seen even in an enlarged image, and it is impossible to distinguish the observed data from the theoretical curve.

The cosmic microwave background radiation is an emission of uniform, black body thermal energy coming from all parts of the sky. The radiation is isotropic to roughly one part in 100,000: the root mean square variations are only 18 μK, after subtracting out a dipole anisotropy from the Doppler shift of the background radiation. The latter is caused by the peculiar velocity of the Sun relative to the comoving cosmic rest frame as it moves at some 369.82±0.11 km/s towards the constellation Leo (galactic longitude 264.021±0.011, galactic latitude 48.253±0.005). The CMB dipole as well as aberration at higher multipoles have been measured, consistent with galactic motion.

In the Big Bang model for the formation of the universe, inflationary cosmology predicts that after about $10^{-37}$ seconds the nascent universe underwent exponential growth that smoothed out nearly all irregularities. The remaining irregularities were caused by quantum fluctuations in the inflaton field that caused the inflation event. Long before the formation of stars and planets, the early universe was smaller, much hotter and, starting $10^{-6}$ seconds after the Big Bang, filled with a uniform glow from its white-hot fog of interacting plasma of photons, electrons, and baryons.

As the universe expanded, adiabatic cooling caused the energy density of the plasma to decrease until it became favorable for electrons to combine with protons, forming hydrogen atoms. This recombination event happened when the temperature was around 3000 K or when the universe was approximately 379,000 years old. As photons did not interact with these electrically neutral atoms, the former began to travel freely through space, resulting in the decoupling of matter and radiation.

The color temperature of the ensemble of decoupled photons has continued to diminish ever since; now down to 2.7260±0.0013 K, it will continue to drop as the universe expands. The intensity of the radiation also corresponds to black-body radiation at 2.726 K because red-shifted black-body radiation is just like black-body radiation at a lower temperature. According to the Big Bang model, the radiation from the sky we measure today comes from a spherical surface called *the surface of last scattering*. This represents the set of locations in space at which the decoupling event is estimated to have occurred and at a point in time such that the photons from that distance have just reached observers. Most of the radiation energy in the universe is in the cosmic microwave background, making up a fraction of roughly $6 \times 10^{-5}$ of the total density of the universe.

Two of the greatest successes of the Big Bang theory are its prediction of the almost perfect black body spectrum and its detailed prediction of the anisotropies in the cosmic microwave background. The CMB spectrum has become the most precisely measured black body spectrum in nature.

Density of energy for CMB is 0.25 eV/cm³ ($4.005 \times 10^{-14}$ J/m³) or (400–500 photons/cm³).

## Relationship to the Big Bang

The cosmic microwave background radiation and the cosmological redshift-distance relation are together regarded as the best available evidence for the Big Bang theory. Measurements of the CMB have made the inflationary Big Bang theory the Standard Cosmological Model. The discovery of the CMB in the mid-1960s curtailed interest in alternatives such as the steady state theory.

The CMB essentially confirms the Big Bang theory. In the late 1940s Alpher and Herman reasoned that if there was a big bang, the expansion of the universe would have stretched and cooled the high-energy radiation of the very early universe into the microwave region of the electromagnetic spectrum, and down to a temperature of about

5 K. They were slightly off with their estimate, but they had exactly the right idea. They predicted the CMB. It took another 15 years for Penzias and Wilson to stumble into discovering that the microwave background was actually there.

The CMB gives a snapshot of the universe when, according to standard cosmology, the temperature dropped enough to allow electrons and protons to form hydrogen atoms, thereby making the universe nearly transparent to radiation because light was no longer being scattered off free electrons. When it originated some 380,000 years after the Big Bang—this time is generally known as the "time of last scattering" or the period of recombination or decoupling—the temperature of the universe was about 3000 K. This corresponds to an energy of about 0.26 eV, which is much less than the 13.6 eV ionization energy of hydrogen.

Since decoupling, the temperature of the background radiation has dropped by a factor of roughly 1,100 due to the expansion of the universe. As the universe expands, the CMB photons are redshifted, causing them to decrease in energy. The temperature of this radiation stays inversely proportional to a parameter that describes the relative expansion of the universe over time, known as the scale length. The temperature $T_r$ of the CMB as a function of redshift, $z$, can be shown to be proportional to the temperature of the CMB as observed in the present day (2.725 K or 0.2348 meV):

$$T_r = 2.725(1 + z)$$

## Primary Anisotropy

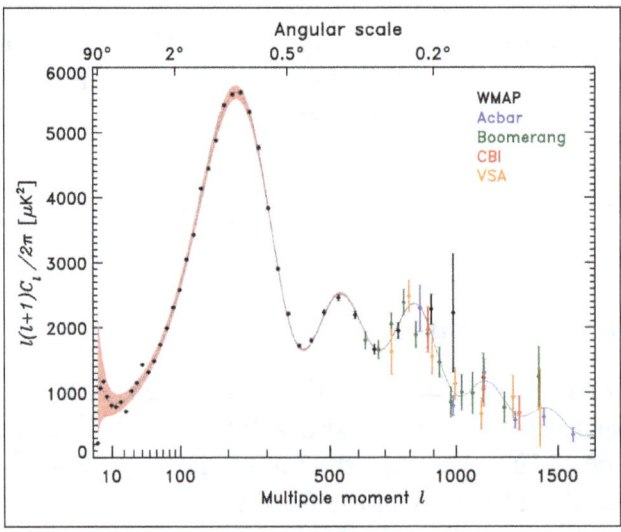

The power spectrum of the cosmic microwave background radiation temperature anisotropy in terms of the angular scale (or multipole moment).

The anisotropy, or directional dependency, of the cosmic microwave background is divided into two types: primary anisotropy, due to effects that occur at the last scattering surface and before; and secondary anisotropy, due to effects such as interactions of the background radiation with hot gas or gravitational potentials, which occur between the last scattering surface and the observer.

The structure of the cosmic microwave background anisotropies is principally determined by two effects: acoustic oscillations and diffusion damping (also called collisionless damping or Silk damping). The acoustic oscillations arise because of a conflict in the photon–baryon plasma in the early universe. The pressure of the photons tends to erase anisotropies, whereas the gravitational attraction of the baryons, moving at speeds much slower than light, makes them tend to collapse to form overdensities. These two effects compete to create acoustic oscillations, which give the microwave background its characteristic peak structure. The peaks correspond, roughly, to resonances in which the photons decouple when a particular mode is at its peak amplitude.

The peaks contain interesting physical signatures. The angular scale of the first peak determines the curvature of the universe (but not the topology of the universe). The next peak—ratio of the odd peaks to the even peaks—determines the reduced baryon density. The third peak can be used to get information about the dark-matter density.

The locations of the peaks also give important information about the nature of the primordial density perturbations. There are two fundamental types of density perturbations called *adiabatic* and *isocurvature*. A general density perturbation is a mixture of both, and different theories that purport to explain the primordial density perturbation spectrum predict different mixtures.

- Adiabatic density perturbations:

  In an adiabatic density perturbation, the fractional additional number density of each type of particle (baryons, photons) is the same. That is, if at one place there is a 1% higher number density of baryons than average, then at that place there is also a 1% higher number density of photons (and a 1% higher number density in neutrinos) than average. Cosmic inflation predicts that the primordial perturbations are adiabatic.

- Isocurvature density perturbations:

  In an isocurvature density perturbation, the sum (over different types of particle) of the fractional additional densities is zero. That is, a perturbation where at some spot there is 1% more energy in baryons than average, 1% more energy in photons than average, and 2% *less* energy in neutrinos than average, would be a pure isocurvature perturbation. Cosmic strings would produce mostly isocurvature primordial perturbations.

The CMB spectrum can distinguish between these two because these two types of perturbations produce different peak locations. Isocurvature density perturbations produce a series of peaks whose angular scales ($l$ values of the peaks) are roughly in the ratio 1:3:5:..., while adiabatic density perturbations produce peaks whose locations are in the ratio 1:2:3:... Observations are consistent with the primordial density perturbations being entirely adiabatic, providing key support for inflation, and ruling out many models of structure formation involving, for example, cosmic strings.

Collisionless damping is caused by two effects, when the treatment of the primordial plasma as fluid begins to break down:

- The increasing mean free path of the photons as the primordial plasma becomes increasingly rarefied in an expanding universe.

- The finite depth of the last scattering surface (LSS), which causes the mean free path to increase rapidly during decoupling, even while some Compton scattering is still occurring.

These effects contribute about equally to the suppression of anisotropies at small scales and give rise to the characteristic exponential damping tail seen in the very small angular scale anisotropies.

The depth of the LSS refers to the fact that the decoupling of the photons and baryons does not happen instantaneously, but instead requires an appreciable fraction of the age of the universe up to that era. One method of quantifying how long this process took uses the *photon visibility function* (PVF). This function is defined so that, denoting the PVF by $P(t)$, the probability that a CMB photon last scattered between time $t$ and $t + dt$ is given by $P(t)\,dt$.

The maximum of the PVF (the time when it is most likely that a given CMB photon last scattered) is known quite precisely. The first-year WMAP results put the time at which $P(t)$ has a maximum as 372,000 years. This is often taken as the "time" at which the CMB formed. However, to figure out how *long* it took the photons and baryons to decouple, we need a measure of the width of the PVF. The WMAP team finds that the PVF is greater than half of its maximal value (the "full width at half maximum", or FWHM) over an interval of 115,000 years. By this measure, decoupling took place over roughly 115,000 years, and when it was complete, the universe was roughly 487,000 years old.

## Late Time Anisotropy

Since the CMB came into existence, it has apparently been modified by several subsequent physical processes, which are collectively referred to as late-time anisotropy, or secondary anisotropy. When the CMB photons became free to travel unimpeded, ordinary matter in the universe was mostly in the form of neutral hydrogen and helium atoms. However, observations of galaxies today seem to indicate that most of the volume of the intergalactic medium (IGM) consists of ionized material (since there are few absorption lines due to hydrogen atoms). This implies a period of reionization during which some of the material of the universe was broken into hydrogen ions.

The CMB photons are scattered by free charges such as electrons that are not bound in atoms. In an ionized universe, such charged particles have been liberated from neutral atoms by ionizing (ultraviolet) radiation. Today these free charges are at sufficiently low density in most of the volume of the universe that they do not measurably affect the CMB. However, if the IGM was ionized at very early times when the universe was still denser, then there are two main effects on the CMB:

- Small scale anisotropies are erased. (Just as when looking at an object through fog, details of the object appear fuzzy.)

- The physics of how photons are scattered by free electrons (Thomson scattering) induces polarization anisotropies on large angular scales. This broad angle polarization is correlated with the broad angle temperature perturbation.

Both of these effects have been observed by the WMAP spacecraft, providing evidence that the universe was ionized at very early times, at a redshift more than 17. The detailed provenance of this early ionizing radiation is still a matter of scientific debate. It may have included starlight from the very first population of stars (population III stars), supernovae when these first stars reached the end of their lives, or the ionizing radiation produced by the accretion disks of massive black holes.

The time following the emission of the cosmic microwave background—and before the observation of the first stars—is semi-humorously referred to by cosmologists as the Dark Age, and is a period which is under intense study by astronomers.

Two other effects which occurred between reionization and our observations of the cosmic microwave background, and which appear to cause anisotropies, are the Sunyaev–Zel'dovich effect, where a cloud of high-energy electrons scatters the radiation, transferring some of its energy to the CMB photons, and the Sachs–Wolfe effect, which causes photons from the Cosmic Microwave Background to be gravitationally redshifted or blueshifted due to changing gravitational fields.

## Polarization

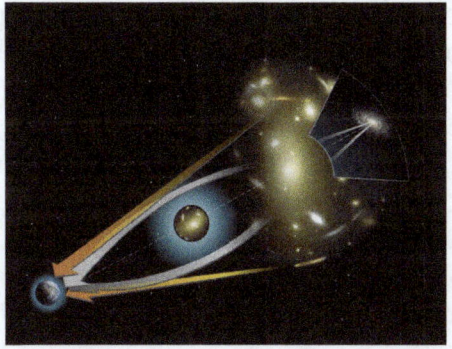

This artist's impression shows how light from the early universe is deflected by the gravitational lensing effect of massive cosmic structures forming B-modes as it travels across the universe.

The cosmic microwave background is polarized at the level of a few microkelvin. There are two types of polarization, called E-modes and B-modes. This is in analogy to electrostatics, in which the electric field (E-field) has a vanishing curl and the magnetic field (B-field) has a vanishing divergence. The E-modes arise naturally from Thomson scattering in a heterogeneous plasma. The B-modes are not produced by standard scalar type perturbations. Instead they can be created by two mechanisms: the first one is by gravitational lensing of E-modes, which has been measured by the South Pole Telescope in 2013; the second one is from gravitational waves arising from cosmic inflation. Detecting the B-modes is extremely difficult, particularly as the degree of foreground contamination is unknown, and the weak gravitational lensing signal mixes the relatively strong E-mode signal with the B-mode signal.

## E-modes

E-modes were first seen in 2002 by the Degree Angular Scale Interferometer (DASI).

## B-modes

Cosmologists predict two types of B-modes, the first generated during cosmic inflation shortly after the big bang, and the second generated by gravitational lensing at later times.

## Primordial Gravitational Waves

Primordial gravitational waves are gravitational waves that could be observed in the polarisation of the cosmic microwave background and having their origin in the early universe. Models of cosmic inflation predict that such gravitational waves should appear; thus, their detection supports the theory of inflation, and their strength can confirm and exclude different models of inflation. It is the result of three things: inflationary expansion of space itself, reheating after inflation, and turbulent fluid mixing of matter and radiation.

On 17 March 2014 it was announced that the BICEP2 instrument had detected the first type of B-modes, consistent with inflation and gravitational waves in the early universe at the level of $r = 0.20+0.07-0.05$, which is the amount of power present in gravitational waves compared to the amount of power present in other scalar density perturbations in the very early universe. Had this been confirmed it would have provided strong evidence of cosmic inflation and the Big Bang, but on 19 June 2014, considerably lowered confidence in confirming the findings was reported and on 19 September 2014 new results of the Planck experiment reported that the results of BICEP2 can be fully attributed to cosmic dust.

## Gravitational Lensing

The second type of B-modes was discovered in 2013 using the South Pole Telescope with help from the Herschel Space Observatory. This discovery may help test theories on the origin of the universe. Scientists are using data from the Planck mission by the European Space Agency, to gain a better understanding of these waves.

In October 2014, a measurement of the B-mode polarization at 150 GHz was published by the POLARBEAR experiment. Compared to BICEP2, POLARBEAR focuses on a smaller patch of the sky and is less susceptible to dust effects. The team reported that POLARBEAR's measured B-mode polarization was of cosmological origin (and not just due to dust) at a 97.2% confidence level.

## Microwave Background Observations

Subsequent to the discovery of the CMB, hundreds of cosmic microwave background experiments have been conducted to measure and characterize the signatures of the radiation. The most famous experiment is probably the NASA Cosmic Background Explorer (COBE) satellite that orbited in 1989–1996 and which detected and quantified the large scale anisotropies at the limit of its detection capabilities. Inspired by the initial COBE results of an extremely isotropic and homogeneous background, a series of ground- and balloon-based experiments quantified CMB anisotropies on smaller angular scales over the next decade. The primary goal of these experiments was to measure the angular scale of the first acoustic peak, for which COBE did not have sufficient resolution. These measurements were able to rule out cosmic strings as the leading theory of cosmic structure formation, and suggested cosmic inflation was the right theory. During the 1990s, the first peak was measured with increasing sensitivity and by 2000 the BOOMERanG experiment reported that the highest power fluctuations occur at scales of approximately one degree. Together with other cosmological data, these results implied that the geometry of the universe is flat. A number of ground-based interferometers provided measurements of the fluctuations with higher accuracy over the next three years, including the Very Small Array, Degree Angular Scale Interferometer (DASI), and the Cosmic Background Imager (CBI). DASI made the first detection of the polarization of the CMB and the CBI provided the first E-mode polarization spectrum with compelling evidence that it is out of phase with the T-mode spectrum.

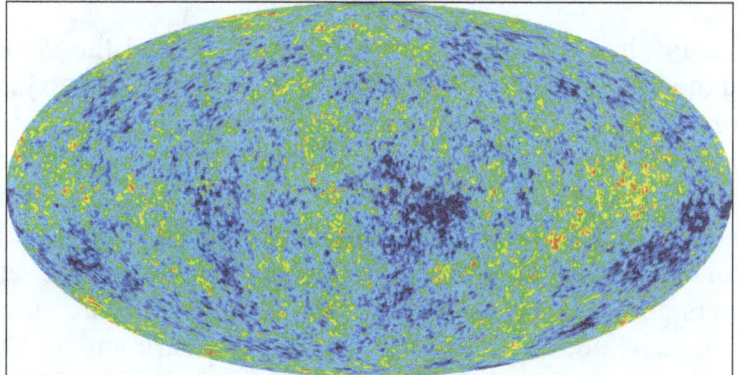

All-sky mollweide map of the CMB, created from 9 years of WMAP data.

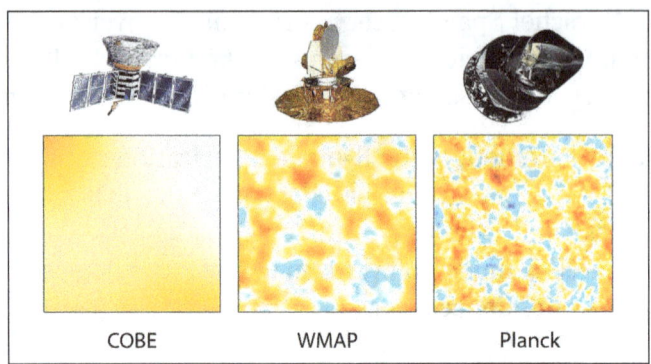

COBE          WMAP          Planck

Comparison of CMB results from COBE, WMAP and Planck.

In June 2001, NASA launched a second CMB space mission, WMAP, to make much more precise measurements of the large scale anisotropies over the full sky. WMAP used symmetric, rapid-multi-modulated scanning, rapid switching radiometers to minimize non-sky signal noise. The first results from this mission, disclosed in 2003, were detailed measurements of the angular power spectrum at a scale of less than one degree, tightly constraining various cosmological parameters. The results are broadly consistent with those expected from cosmic inflation as well as various other competing theories, and are available in detail at NASA's data bank for Cosmic Microwave Background (CMB). Although WMAP provided very accurate measurements of the large scale angular fluctuations in the CMB (structures about as broad in the sky as the moon), it did not have the angular resolution to measure the smaller scale fluctuations which had been observed by former ground-based interferometers.

A third space mission, the ESA (European Space Agency) Planck Surveyor, was launched in May 2009 and performed an even more detailed investigation until it was shut down in October 2013. Planck employed both HEMT radiometers and bolometer technology and measured the CMB at a smaller scale than WMAP. Its detectors were trialled in the Antarctic Viper telescope as ACBAR (Arcminute Cosmology Bolometer Array Receiver) experiment—which has produced the most precise measurements at small angular scales to date—and in the Archeops balloon telescope.

On 21 March 2013, the European-led research team behind the Planck cosmology probe released the mission's all-sky map (565x318 jpeg, 3600x1800 jpeg) of the cosmic microwave background. The map suggests the universe is slightly older than researchers expected. According to the map, subtle fluctuations in temperature were imprinted on the deep sky when the cosmos was about 370000 years old. The imprint reflects ripples that arose as early, in the existence of the universe, as the first nonillionth of a second. Apparently, these ripples gave rise to the present vast cosmic web of galaxy clusters and dark matter. Based on the 2013 data, the universe contains 4.9% ordinary matter, 26.8% dark matter and 68.3% dark energy. On 5 February 2015, new data was released by the Planck mission, according to which the age of the universe is 13.799±0.021 billion years old and the Hubble constant was measured to be 67.74±0.46 (km/s)/Mpc.

Additional ground-based instruments such as the South Pole Telescope in Antarctica and the proposed Clover Project, Atacama Cosmology Telescope and the QUIET tele-scope in Chile will provide additional data not available from satellite observations, possibly including the B-mode polarization.

## Data Reduction and Analysis

Raw CMBR data, even from space vehicles such as WMAP or Planck, contain fore-ground effects that completely obscure the fine-scale structure of the cosmic micro-wave background. The fine-scale structure is superimposed on the raw CMBR data but is too small to be seen at the scale of the raw data. The most prominent of the fore-ground effects is the dipole anisotropy caused by the Sun's motion relative to the CMBR background. The dipole anisotropy and others due to Earth's annual motion relative to the Sun and numerous microwave sources in the galactic plane and elsewhere must be subtracted out to reveal the extremely tiny variations characterizing the fine-scale structure of the CMBR background.

The detailed analysis of CMBR data to produce maps, an angular power spectrum, and ultimately cosmological parameters is a complicated, computationally difficult problem. Although computing a power spectrum from a map is in principle a sim-ple Fourier transform, decomposing the map of the sky into spherical harmonics, in practice it is hard to take the effects of noise and foreground sources into account. In particular, these foregrounds are dominated by galactic emissions such as Brems-strahlung, synchrotron, and dust that emit in the microwave band; in practice, the galaxy has to be removed, resulting in a CMB map that is not a full-sky map. In addi-tion, point sources like galaxies and clusters represent another source of foreground which must be removed so as not to distort the short scale structure of the CMB pow-er spectrum.

Constraints on many cosmological parameters can be obtained from their effects on the power spectrum, and results are often calculated using Markov chain Monte Carlo sampling techniques.

## CMBR Dipole Anisotropy

From the CMB data it is seen that the earth appears to be moving at 368±2 km/s relative to the reference frame of the CMB (also called the CMB rest frame, or the frame of reference in which there is no motion through the CMB). The Local Group (the galaxy group that includes the Milky Way galaxy) appears to be moving at 627±22 km/s in the direction of galactic longitude $l = 276°±3°$, $b = 30°±3°$. This motion results in an anisotropy of the data (CMB appearing slightly warmer in the direction of movement than in the opposite direction). From a theoretical point of view, the existence of a CMB rest frame breaks Lorentz invariance even in empty space far away from any galaxy. The standard interpretation of this temperature variation is a simple velocity red shift and blue shift due to motion relative to the CMB, but alternative cosmological models can explain some fraction of the observed dipole temperature distribution in the CMB.

## Low Multipoles and other Anomalies

With the increasingly precise data provided by WMAP, there have been a number of claims that the CMB exhibits anomalies, such as very large scale anisotropies, anomalous alignments, and non-Gaussian distributions. The most longstanding of these is the low-$l$ multipole controversy. Even in the COBE map, it was observed that the quadrupole ($l = 2$, spherical harmonic) has a low amplitude compared to the predictions of the Big Bang. In particular, the quadrupole and octupole ($l = 3$) modes appear to have an unexplained alignment with each other and with both the ecliptic plane and equinoxes, A number of groups have suggested that this could be the signature of new physics at the greatest observable scales; other groups suspect systematic errors in the data. Ultimately, due to the foregrounds and the cosmic variance problem, the greatest modes will never be as well measured as the small angular scale modes. The analyses were performed on two maps that have had the foregrounds removed as far as possible: the "internal linear combination" map of the WMAP collaboration and a similar map prepared by Max Tegmark and others. Later analyses have pointed out that these are the modes most susceptible to foreground contamination from synchrotron, dust, and Bremsstrahlung emission, and from experimental uncertainty in the monopole and dipole. A full Bayesian analysis of the WMAP power spectrum demonstrates that the quadrupole prediction of Lambda-CDM cosmology is consistent with the data at the 10% level and that the observed octupole is not remarkable. Carefully accounting for the procedure used to remove the foregrounds from the full sky map further reduces the significance of the alignment by ~5%. Recent observations with the Planck telescope, which is very much more sensitive than WMAP and has a larger angular resolution, record the same anomaly, and so instrumental error (but not foreground contamination) appears to be ruled out. Coincidence is a possible explanation, chief scientist from WMAP, Charles L. Bennett suggested coincidence and human psychology were involved, "I do think there is a bit of a psychological effect; people want to find unusual things."

## Future Evolution

Assuming the universe keeps expanding and it does not suffer a Big Crunch, a Big Rip, or another similar fate, the cosmic microwave background will continue redshifting until it will no longer be detectable, and will be overtaken first by the one produced by starlight, and later by the background radiation fields of processes that are assumed will take place in the far future of the universe.

## Timeline of Prediction, Discovery and Interpretation

### Thermal (Non-microwave Background) Temperature Predictions

- 1896 – Charles Édouard Guillaume estimates the "radiation of the stars" to be 5–6K.

- 1926 – Sir Arthur Eddington estimates the non-thermal radiation of starlight in the galaxy "by the formula $E = \sigma T^4$ the effective temperature corresponding to this density is 3.18° absolute black body".

- 1930s – Cosmologist Erich Regener calculates that the non-thermal spectrum of cosmic rays in the galaxy has an effective temperature of 2.8 K.

- 1931 – Term *microwave* first used in print: "When trials with wavelengths as low as 18 cm. were made known, there was undisguised surprise+that the problem of the micro-wave had been solved so soon." *Telegraph & Telephone Journal* XVII. 179/1.

- 1934 – Richard Tolman shows that black-body radiation in an expanding universe cools but remains thermal.

- 1938 – Nobel Prize winner (1920) Walther Nernst reestimates the cosmic ray temperature as 0.75K.

- 1946 – Robert Dicke predicts "radiation from cosmic matter" at <20 K, but did not refer to background radiation.

- 1946 – George Gamow calculates a temperature of 50 K (assuming a 3-billion year old universe), commenting it "is in reasonable agreement with the actual temperature of interstellar space", but does not mention background radiation.

- 1953 – Erwin Finlay-Freundlich in support of his tired light theory, derives a blackbody temperature for intergalactic space of 2.3K with comment from Max Born suggesting radio astronomy as the arbitrator between expanding and infinite cosmologies.

### Microwave Background Radiation Predictions and Measurements

- 1941 – Andrew McKellar detected the cosmic microwave background as the coldest component of the interstellar medium by using the excitation of CN doublet lines measured by W. S. Adams in a B star, finding an "effective temperature of space" (the average bolometric temperature) of 2.3 K.

- 1946 – George Gamow calculates a temperature of 50 K (assuming a 3-billion year old universe), commenting it "is in reasonable agreement with the actual temperature of interstellar space", but does not mention background radiation.

- 1948 – Ralph Alpher and Robert Herman estimate "the temperature in the universe" at 5 K. Although they do not specifically mention microwave background radiation, it may be inferred.

- 1949 – Ralph Alpher and Robert Herman re-re-estimate the temperature at 28 K.

- 1953 – George Gamow estimates 7 K.

- 1956 – George Gamow estimates 6 K.

- 1955 – Émile Le Roux of the Nançay Radio Observatory, in a sky survey at $\lambda$ = 33 cm, reported a near-isotropic background radiation of 3 kelvins, plus or minus 2.

- 1957 – Tigran Shmaonov reports that "the absolute effective temperature of the radioemission background is 4±3 K". It is noted that the "measurements showed that radiation intensity was independent of either time or direction of observation it is now clear that Shmaonov did observe the cosmic microwave background at a wavelength of 3.2 cm".

- 1960s – Robert Dicke re-estimates a microwave background radiation temperature of 40 K.

- 1964 – A. G. Doroshkevich and Igor Dmitrievich Novikov publish a brief paper suggesting microwave searches for the black-body radiation predicted by Gamow, Alpher, and Herman, where they name the CMB radiation phenomenon as detectable.

- 1964–65 – Arno Penzias and Robert Woodrow Wilson measure the temperature to be approximately 3 K. Robert Dicke, James Peebles, P. G. Roll, and D. T. Wilkinson interpret this radiation as a signature of the big bang.

- 1966 – Rainer K. Sachs and Arthur M. Wolfe theoretically predict microwave background fluctuation amplitudes created by gravitational potential variations between observers and the last scattering surface 1968 – Martin Rees and Dennis Sciama theoretically predict microwave background fluctuation amplitudes created by photons traversing time-dependent potential wells.

- 1969 – R. A. Sunyaev and Yakov Zel'dovich study the inverse Compton scattering of microwave background photons by hot electrons.

- 1983 – Researchers from the Cambridge Radio Astronomy Group and the Owens Valley Radio Observatory first detect the Sunyaev-Zel'dovich effect from clusters of galaxies.

- 1983 – RELIKT-1 Soviet CMB anisotropy experiment was launched.

- 1990 – FIRAS on the Cosmic Background Explorer (COBE) satellite measures the black body form of the CMB spectrum with exquisite precision, and shows that the microwave background has a nearly perfect black-body spectrum and thereby strongly constrains the density of the intergalactic medium.

- January 1992 – Scientists that analysed data from the RELIKT-1 report the discovery of anisotropy in the cosmic microwave background at the Moscow astrophysical seminar.

- 1992 – Scientists that analysed data from COBE DMR report the discovery of anisotropy in the cosmic microwave background.

- 1995 – The Cosmic Anisotropy Telescope performs the first high resolution observations of the cosmic microwave background.

- 1999 – First measurements of acoustic oscillations in the CMB anisotropy angular power spectrum from the TOCO, BOOMERANG, and Maxima Experiments. The BOOMERanG experiment makes higher quality maps at intermediate resolution, and confirms that the universe is "flat".

- 2002 – Polarization discovered by DASI.

- 2003 – E-mode polarization spectrum obtained by the CBI. The CBI and the Very Small Array produces yet higher quality maps at high resolution (covering small areas of the sky).

- 2003 – The Wilkinson Microwave Anisotropy Probe spacecraft produces an even higher quality map at low and intermediate resolution of the whole sky (WMAP provides *no* high-resolution data, but improves on the intermediate resolution maps from BOOMERanG).

- 2004 – E-mode polarization spectrum obtained by the CBI.

- 2004 – The Arcminute Cosmology Bolometer Array Receiver produces a higher quality map of the high resolution structure not mapped by WMAP.

- 2005 – The Arcminute Microkelvin Imager and the Sunyaev-Zel'dovich Array begin the first surveys for very high redshift clusters of galaxies using the Sunyaev-Zel'dovich effect.

- 2005 – Ralph A. Alpher is awarded the National Medal of Science for his groundbreaking work in nucleosynthesis and prediction that the universe expansion leaves behind background radiation, thus providing a model for the Big Bang theory.

- 2006 – The long-awaited three-year WMAP results are released, confirming previous analysis, correcting several points, and including polarization data.

- 2006 – Two of COBE's principal investigators, George Smoot and John Mather, received the Nobel Prize in Physics in 2006 for their work on precision measurement of the CMBR.

- 2006–2011 – Improved measurements from WMAP, new supernova surveys ESSENCE and SNLS, and baryon acoustic oscillations from SDSS and WiggleZ, continue to be consistent with the standard Lambda-CDM model.

- 2010 – The first all-sky map from the Planck telescope is released.

- 2013 – An improved all-sky map from the Planck telescope is released, improving the measurements of WMAP and extending them to much smaller scales.

- 2014 – On March 17, 2014, astrophysicists of the BICEP2 collaboration announced the detection of inflationary gravitational waves in the B-mode power spectrum, which if confirmed, would provide clear experimental evidence for the theory of inflation. However, on 19 June 2014, lowered confidence in confirming the cosmic inflation findings was reported.

- 2015 – On January 30, 2015, the same team of astronomers from BICEP2 withdrew the claim made on the previous year. Based on the combined data of BICEP2 and Planck, the European Space Agency announced that the signal can be entirely attributed to dust in the Milky Way.

- 2018 – The final data and maps from the Planck telescope is released, with improved measurements of the polarization on large scales.

## Luminosity

Luminosity is an absolute measure of radiated electromagnetic power (light), the radiant power emitted by a light-emitting object.

In astronomy, luminosity is the total amount of electromagnetic energy emitted per unit of time by a star, galaxy, or other astronomical object.

In SI units, luminosity is measured in joules per second, or watts. In astronomy, values for luminosity are often given in the terms of the luminosity of the Sun, $L_\odot$. Luminosity can also be given in terms of the astronomical magnitude system: the absolute bolometric magnitude ($M_{bol}$) of an object is a logarithmic measure of its total energy emission rate, while absolute magnitude is a logarithmic measure of the luminosity within some specific wavelength range or filter band.

In contrast, the term *brightness* in astronomy is generally used to refer to an object's apparent brightness: that is, how bright an object appears to an observer. Apparent brightness depends on both the luminosity of the object and the distance between the object and observer, and also on any absorption of light along the path from object to observer. Apparent magnitude is a logarithmic measure of apparent brightness. The distance determined by luminosity measures can be somewhat ambiguous, and is thus sometimes called the luminosity distance.

The sun has an intrinsic luminosity of $3.83 \times 10^{26}$ Watts. In astronomy, this amount is equal to one solar luminosity, represented by the symbol $L_\odot$. A star with four times the radiative power of the sun has a luminosity of $4\,L_\odot$.

## Measurement

When not qualified, the term "luminosity" means bolometric luminosity, which is measured either in the SI units, watts, or in terms of solar luminosities ($L_\odot$). A bolometer is the instrument used to measure radiant energy over a wide band by absorption and measurement of heating. A star also radiates neutrinos, which carry off some energy (about 2% in the case of our Sun), contributing to the star's total luminosity. The IAU has defined a nominal solar luminosity of $3.828 \times 10^{26}$ W to promote publication of consistent and comparable values in units of the solar luminosity.

While bolometers do exist, they cannot be used to measure even the apparent brightness of a star because they are insufficiently sensitive across the electromagnetic spectrum and because most wavelengths do not reach the surface of the Earth. In practice bolometric magnitudes are measured by taking measurements at certain wavelengths and constructing a model of the total spectrum that is most likely to match those measurements. In some cases, the process of estimation is extreme, with luminosities being calculated when less than 1% of the energy output is observed, for example with a hot Wolf-Rayet star observed only in the infra-red. Bolometric luminosities can also be calculated using a bolometric correction to a luminosity in a particular passband.

The term luminosity is also used in relation to particular passbands such as a visual luminosity of K-band luminosity. These are not generally luminosities in the strict sense of an absolute measure of radiated power, but absolute magnitudes defined for a given filter in a photometric system. Several different photometric systems exist. Some such as the UBV or Johnson system are defined against photometric standard stars, while others such as the AB system are defined in terms of a spectral flux density.

## Stellar Luminosity

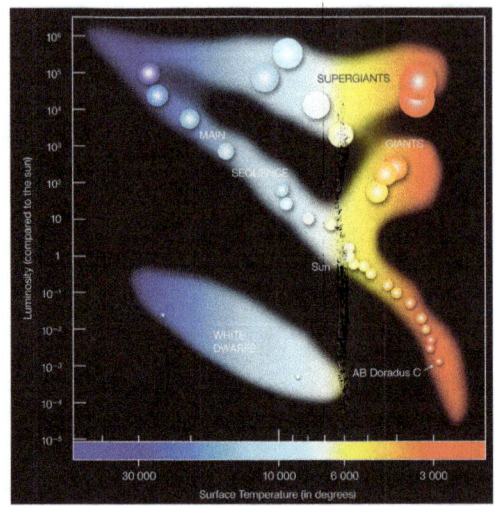

Hertzsprung–Russell diagram identifying stellar luminosity as a function of temperature for many stars in our solar neighborhood.

A star's luminosity can be determined from two stellar characteristics: size and effective temperature. The former is typically represented in terms of solar radii, $R_\odot$, while the latter is represented in kelvins, but in most cases neither can be measured directly. To determine a star's radius, two other metrics are needed: the star's angular diameter and its distance from Earth. Both can be measured with great accuracy in certain cases, with cool supergiants often having large angular diameters, and some cool evolved stars having masers in their atmospheres that can be used to measure the parallax using VLBI. However, for most stars the angular diameter or parallax, or both, are far below our ability to measure with any certainty. Since the effective temperature is merely a number that represents the temperature of a black body that would reproduce the luminosity, it obviously cannot be measured directly, but it can be estimated from the spectrum.

An alternative way to measure stellar luminosity is to measure the star's apparent brightness and distance. A third component needed to derive the luminosity is the degree of interstellar extinction that is present, a condition that usually arises because of gas and dust present in the interstellar medium (ISM), the Earth's atmosphere, and circumstellar matter. Consequently, one of astronomy's central challenges in determining a star's luminosity is to derive accurate measurements for each of these components, without which an accurate luminosity figure remains elusive. Extinction can only be measured directly if the actual and observed luminosities are both known, but it can be estimated from the observed colour of a star, using models of the expected level of reddening from the interstellar medium.

In the current system of stellar classification, stars are grouped according to temperature, with the massive, very young and energetic Class O stars boasting temperatures in excess of 30,000 K while the less massive, typically older Class M stars exhibit temperatures less than 3,500 K. Because luminosity is proportional to temperature to the fourth power, the large variation in stellar temperatures produces an even vaster variation in stellar luminosity. Because the luminosity depends on a high power of the stellar mass, high mass luminous stars have much shorter lifetimes. The most luminous stars are always young stars, no more than a few million years for the most extreme. In the Hertzsprung–Russell diagram, the x-axis represents temperature or spectral type while the y-axis represents luminosity or magnitude. The vast majority of stars are found along the main sequence with blue Class O stars found at the top left of the chart while red Class M stars fall to the bottom right. Certain stars like Deneb and Betelgeuse are found above and to the right of the main sequence, more luminous or cooler than their equivalents on the main sequence. Increased luminosity at the same temperature, or alternatively cooler temperature at the same luminosity, indicates that these stars are larger than those on the main sequence and they are called giants or supergiants.

Blue and white supergiants are high luminosity stars somewhat cooler than the most luminous main sequence stars. A star like Deneb, for example, has a luminosity around 200,000 $L_\odot$, a spectral type of A2, and an effective temperature around 8,500 K, meaning it has a radius around 203 $R_\odot$. For comparison, the red supergiant Betelgeuse has a luminosity around 100,000 $L_\odot$, a spectral type of M2, and a temperature around 3,500 K, meaning its radius is about 1,000 $R_\odot$. Red supergiants are the largest type of star, but the most luminous are much smaller and hotter, with temperatures up to 50,000 K and more and luminosities of several million $L_\odot$, meaning their radii are just a few tens of $R_\odot$. For example, R136a1 has a temperature over 50,000 K and a luminosity of more than 8,000,000 $L_\odot$ (mostly in the UV), it is only 35 $R_\odot$.

## Radio Luminosity

The luminosity of a radio source is measured in W Hz$^{-1}$, to avoid having to specify a bandwidth over which it is measured. The observed strength, or flux density, of a radio source is measured in Jansky where 1 Jy = $10^{-26}$ W m$^{-2}$ Hz$^{-1}$.

For example, consider a 10W transmitter at a distance of 1 million metres, radiating over a bandwidth of 1 MHz. By the time that power has reached the observer, the power is spread over the surface of a sphere with area $4\pi r^2$ or about $1.26\times10^{13}$ m$^2$, so its flux density is 10 / $10^6$ / $1.26\times10^{13}$ W m$^{-2}$ Hz$^{-1}$ = $10^8$ Jy.

More generally, for sources at cosmological distances, a k-correction must be made for the spectral index $\alpha$ of the source, and a relativistic correction must be made for the fact that the frequency scale in the emitted rest frame is different from that in the observer's rest frame. So the full expression for radio luminosity, assuming isotropic emission, is:

$$L_\nu = \frac{S_{obs} 4\pi D_L{}^2}{(1+z)^{1+\alpha}}$$

where $L_\nu$ is the luminosity in W Hz$^{-1}$, $S_{obs}$ is the observed flux density in W m$^{-2}$ Hz$^{-1}$, $D_L$ is the luminosity distance in metres, $z$ is the redshift, $\alpha$ is the spectral index (in the sense $I \propto \nu^\alpha$, and in radio astronomy, assuming thermal emission the spectral index is typically equal to 2.)

For example, consider a 1 Jy signal from a radio source at a redshift of 1, at a frequency of 1.4 GHz. Ned Wright's cosmology calculator calculates a luminosity distance for a redshift of 1 to be 6701 Mpc = $2\times10^{26}$ m giving a radio luminosity of $10^{-26} \times 4\pi(2\times10^{26})^2$ / $(1+1)^{(1+2)}$ = $6\times10^{26}$ W Hz$^{-1}$.

To calculate the total radio power, this luminosity must be integrated over the bandwidth of the emission. A common assumption is to set the bandwidth to the observing frequency, which effectively assumes the power radiated has uniform intensity from zero frequency up to the observing frequency. In the case above, the total power is $4\times10^{27} \times 1.4\times10^9$ = $5.7\times10^{36}$ W. This is sometimes expressed in terms of the total (i.e. integrated over all wavelengths) luminosity of the Sun which is $3.86\times10^{26}$ W, giving a radio power of $1.5\times10^{10}$ L$_\odot$.

## Luminosity Formulae

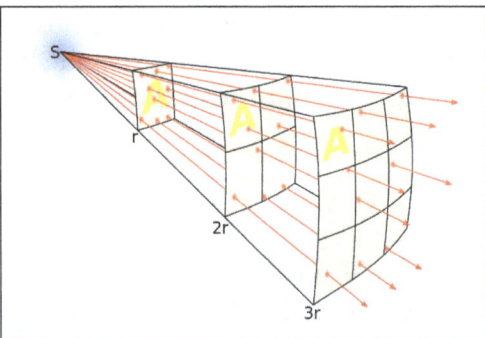

Point source $S$ is radiating light equally in all directions. The amount passing through an area $A$ varies with the distance of the surface from the light.

The Stefan–Boltzmann equation applied to a black body gives the value for luminosity for a black body, an idealized object which is perfectly opaque and non-reflecting:

$$L = \sigma A T^4,$$

where A is the surface area, T is the temperature (in Kelvins) and σ is the Stefan–Boltzmann constant, with a value of 5.670374419...×10$^{-8}$ W·m$^{-2}$·K$^{-4}$.

Imagine a point source of light of luminosity $L$ that radiates equally in all directions. A hollow sphere centered on the point would have its entire interior surface illuminated. As the radius increases, the surface area will also increase, and the constant luminosity has more surface area to illuminate, leading to a decrease in observed brightness.

$$F = \frac{L}{A},$$

where,

$A$ is the area of the illuminated surface.

$F$ is the flux density of the illuminated surface.

The surface area of a sphere with radius $r$ is $A = 4\pi r^2$, so for stars and other point sources of light:

$$F = \frac{L}{4\pi r^2},$$

where, $r$ is the distance from the observer to the light source.

For stars on the main sequence, luminosity is also related to mass:

$$\frac{L}{L_\odot} \approx \left(\frac{M}{M_\odot}\right)^{3.5}.$$

## Relationship to Magnitude

Luminosity is an intrinsic measurable property of a star independent of distance. The concept of magnitude, on the other hand, incorporates distance. The apparent magnitude is a measure of the diminishing flux of light as a result of distance according to the inverse-square law. The Pogson logarithmic scale is used to measure both apparent and absolute magnitudes, the latter corresponding to the brightness of a star or other celestial body as seen if it would be located at an interstellar distance of 10 parsecs .In addition to this brightness decrease from increased distance, there is an extra decrease of brightness due to extinction from intervening interstellar dust.

By measuring the width of certain absorption lines in the stellar spectrum, it is often possible to assign a certain luminosity class to a star without knowing its distance. Thus a fair measure of its absolute magnitude can be determined without knowing its distance nor the interstellar extinction.

In measuring star brightnesses, absolute magnitude, apparent magnitude, and distance are interrelated parameters—if two are known, the third can be determined. Since the

Sun's luminosity is the standard, comparing these parameters with the Sun's apparent magnitude and distance is the easiest way to remember how to convert between them, although officially, zero point values are defined by the IAU.

The magnitude of a star, a unitless measure, is a logarithmic scale of observed visible brightness. The apparent magnitude is the observed visible brightness from Earth which depends on the distance of the object. The absolute magnitude is the apparent magnitude at a distance of 10 parsecs, therefore the bolometric absolute magnitude is a logarithmic measure of the bolometric luminosity.

The difference in bolometric magnitude between two objects is related to their luminosity ratio according to:

$$M_{bol1} - M_{bol2} = -2.5 \log_{10} \frac{L_1}{L_2}$$

where,

- $M_{bol1}$ is the bolometric magnitude of the first object.

- $M_{bol2}$ is the bolometric magnitude of the second object.

- $L_1$ is the first object's bolometric luminosity.

- $L_2$ is the second object's bolometric luminosity.

The zero point of the absolute magnitude scale is actually defined as a fixed luminosity of $3.0128 \times 10^{28}$ W. Therefore, the absolute magnitude can be calculated from a luminosity in watts:

$$M_{bol} = -2.5 \log_{10} \frac{L_*}{L_0} \approx -2.5 \log_{10} L_* + 71.1974$$

where $L_0$ is the zero point luminosity $3.0128 \times 10^{28}$ W

and the luminosity in watts can be calculated from an absolute magnitude (although absolute magnitudes are often not measured relative to an absolute flux):

$$L_* = L_0 \times 10^{-0.4 M_{bol}}.$$

## Dark Matter

Dark matter is a form of matter thought to account for approximately 85% of the matter in the universe and about a quarter of its total energy density. The majority of dark matter is thought to be non-baryonic in nature, possibly being composed of some as-yet undiscovered subatomic particles. Its presence is implied in a variety of astrophysical observations, including gravitational effects which cannot be explained by accepted theories of gravity unless more matter is present than can be seen. For this reason, most experts think dark matter to be abundant in the universe and to have had a strong influence on its structure and evolution. Dark matter is called dark because it does not appear to interact with observable electromagnetic radiation, such as light, and is thus invisible to the entire electromagnetic spectrum, making it extremely difficult to detect using usual astronomical equipment.

Primary evidence for dark matter comes from calculations showing many galaxies would fly apart instead of rotating, or would not have formed or move as they do, if they did not contain a large amount of unseen matter. Other lines of evidence include observations in gravitational lensing, from the cosmic microwave background, also astronomical observations of the observable universe's current structure, the formation and evolution of galaxies, mass location during galactic collisions, and the motion of galaxies within galaxy clusters. In the standard Lambda-CDM model of cosmology, the total mass–energy of the universe contains 5% ordinary matter and energy, 27% dark matter and 68% of an unknown form of energy known as dark energy. Thus, dark matter constitutes 85% of total mass, while dark energy plus dark matter constitute 95% of total mass–energy content.

Because dark matter has not yet been observed directly, if it exists, it must barely interact with ordinary baryonic matter and radiation, except through gravity. The primary candidate for dark matter is some new kind of elementary particle that has not yet been discovered, in particular, weakly-interacting massive particles (WIMPs). Many experiments to directly detect and study dark matter particles are being actively undertaken, but none have yet succeeded. Dark matter is classified as "cold", "warm", or "hot" according to its velocity (more precisely, its free streaming length). Current models favor a cold dark matter scenario, in which structures emerge by gradual accumulation of particles.

Although the existence of dark matter is generally accepted by the scientific community, some astrophysicists, intrigued by certain observations which do not fit the dark matter theory, argue for various modifications of the standard laws of general relativity, such as modified Newtonian dynamics, tensor–vector–scalar gravity, or entropic gravity. These models attempt to account for all observations without invoking supplemental non-baryonic matter.

In standard cosmology, matter is anything whose energy density scales with the inverse cube of the scale factor, i.e., $\rho \propto a^{-3}$. This is in contrast to radiation, which scales as the inverse fourth power of the scale factor $\rho \propto a^{-4}$, and a cosmological constant, which is independent of $a$. These scalings can be understood intuitively: for an ordinary particle in a cubical box, doubling the length of the sides of the box decreases the density (and hence energy density) by a factor of eight (= $2^3$). For radiation, the decrease in energy density is larger because an increase in scale factor causes a proportional redshift. A cosmological constant, as an intrinsic property of space, has a constant energy density regardless of the volume under consideration.

In principle, "dark matter" means all components of the universe which are not visible but still obey $\rho \propto a^{-3}$. In practice, the term "dark matter" is often used to mean only the non-baryonic component of dark matter, i.e., excluding "missing baryons." Context will usually indicate which meaning is intended.

## Observational Evidence

### Galaxy Rotation Curves

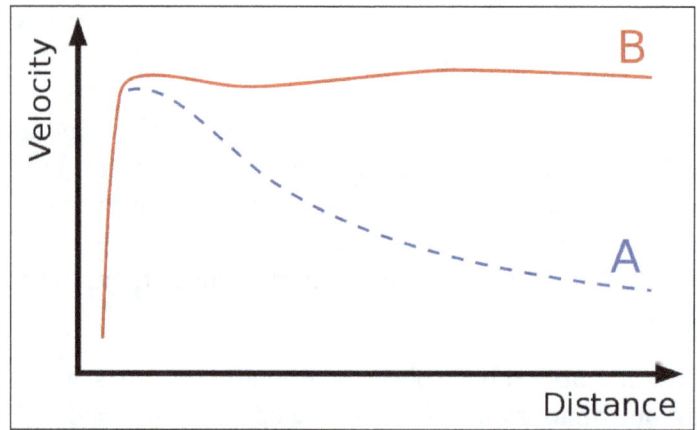

Rotation curve of a typical spiral galaxy: predicted (**A**) and observed (**B**). Dark matter can explain the 'flat' appearance of the velocity curve out to a large radius.

The arms of spiral galaxies rotate around the galactic center. The luminous mass density of a spiral galaxy decreases as one goes from the center to the outskirts. If luminous mass were all the matter, then we can model the galaxy as a point mass in the centre and test masses orbiting around it, similar to the Solar System. From Kepler's Second Law, it is expected that the rotation velocities will decrease with distance from the center, similar to the Solar System. This is not observed. Instead, the galaxy rotation curve remains flat as distance from the center increases.

If Kepler's laws are correct, then the obvious way to resolve this discrepancy is to conclude the mass distribution in spiral galaxies is not similar to that of the Solar System. In particular, there is a lot of non-luminous matter (dark matter) in the outskirts of the galaxy.

### Velocity Dispersions

Stars in bound systems must obey the virial theorem. The theorem, together with the measured velocity distribution, can be used to measure the mass distribution in a bound system, such as elliptical galaxies or globular clusters. With some exceptions,

CHAPTER 3    Basic Concepts in Astrophysics | **119**

velocity dispersion estimates of elliptical galaxies do not match the predicted velocity dispersion from the observed mass distribution, even assuming complicated distributions of stellar orbits.

As with galaxy rotation curves, the obvious way to resolve the discrepancy is to postulate the existence of non-luminous matter.

## Galaxy Clusters

Galaxy clusters are particularly important for dark matter studies since their masses can be estimated in three independent ways:

- From the scatter in radial velocities of the galaxies within clusters.

- From X-rays emitted by hot gas in the clusters. From the X-ray energy spectrum and flux, the gas temperature and density can be estimated, hence giving the pressure; assuming pressure and gravity balance determines the cluster's mass profile.

- Gravitational lensing (usually of more distant galaxies) can measure cluster masses without relying on observations of dynamics (e.g., velocity).

- Generally, these three methods are in reasonable agreement dark matter outweighs visible matter by approximately 5 to 1.

## Gravitational Lensing

Strong gravitational lensing as observed by the Hubble Space Telescope in Abell 1689 indicates the presence of dark matter.

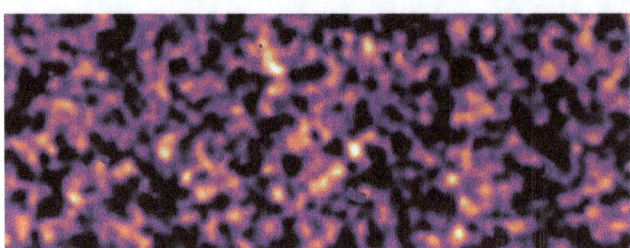

Dark matter map for a patch of sky based on gravitational lensing analysis of
a Kilo-Degree survey.

One of the consequences of general relativity is massive objects (such as a cluster of galaxies) lying between a more distant source (such as a quasar) and an observer should act as a lens to bend the light from this source. The more massive an object, the more lensing is observed.

Strong lensing is the observed distortion of background galaxies into arcs when their light passes through such a gravitational lens. It has been observed around many distant clusters including Abell 1689. By measuring the distortion geometry, the mass of the intervening cluster can be obtained. In the dozens of cases where this has been done, the mass-to-light ratios obtained correspond to the dynamical dark matter measurements of clusters. Lensing can lead to multiple copies of an image. By analyzing the distribution of multiple image copies, scientists have been able to deduce and map the distribution of dark matter around the MACS J0416.1-2403 galaxy cluster.

Weak gravitational lensing investigates minute distortions of galaxies, using statistical analyses from vast galaxy surveys. By examining the apparent shear deformation of the adjacent background galaxies, the mean distribution of dark matter can be characterized. The mass-to-light ratios correspond to dark matter densities predicted by other large-scale structure measurements. Dark matter does not bend light itself; mass (in this case the mass of the dark matter) bends spacetime. Light follows the curvature of spacetime, resulting in the lensing effect.

## Cosmic Microwave Background

Although both dark matter and ordinary matter are matter, they do not behave in the same way. In particular, in the early universe, ordinary matter was ionized and interacted strongly with radiation via Thomson scattering. Dark matter does not interact directly with radiation, but it does affect the CMB by its gravitational potential (mainly on large scales), and by its effects on the density and velocity of ordinary matter. Ordinary and dark matter perturbations, therefore, evolve differently with time and leave different imprints on the cosmic microwave background (CMB).

The cosmic microwave background is very close to a perfect blackbody but contains very small temperature anisotropies of a few parts in 100,000. A sky map of anisotropies can be decomposed into an angular power spectrum, which is observed to contain a series of acoustic peaks at near-equal spacing but different heights. The series of peaks can be predicted for any assumed set of cosmological parameters by modern computer codes such as CMBFast and CAMB, and matching theory to data, therefore, constrains cosmological parameters. The first peak mostly shows the density of baryonic matter, while the third peak relates mostly to the density of dark matter, measuring the density of matter and the density of atoms.

The CMB anisotropy was first discovered by COBE in 1992, though this had too coarse resolution to detect the acoustic peaks. After the discovery of the first acoustic peak by the balloon-borne BOOMERanG experiment in 2000, the power spectrum was precisely observed by WMAP in 2003–2012, and even more precisely by the Planck spacecraft in 2013–2015. The results support the Lambda-CDM model.

The observed CMB angular power spectrum provides powerful evidence in support of dark matter, as its precise structure is well fitted by the Lambda-CDM model, but difficult to reproduce with any competing model such as modified Newtonian dynamics (MOND).

## Structure Formation

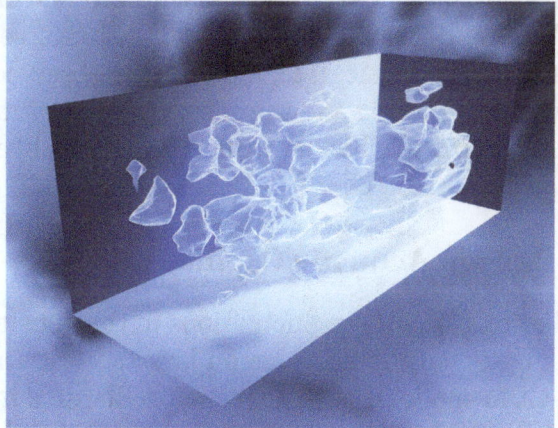

3-D map of the large-scale distribution of dark matter, reconstructed from measurements of weak gravitational lensing with the Hubble Space Telescope.

Structure formation refers to the period after the Big Bang when density perturbations collapsed to form stars, galaxies, and clusters. Prior to structure formation, the Friedmann solutions to general relativity describe a homogeneous universe. Later, small anisotropies gradually grew and condensed the homogeneous universe into stars, galaxies and larger structures. Ordinary matter is affected by radiation, which is the dominant element of the universe at very early times. As a result, its density perturbations are washed out and unable to condense into structure. If there were only ordinary matter in the universe, there would not have been enough time for density perturbations to grow into the galaxies and clusters currently seen.

Dark matter provides a solution to this problem because it is unaffected by radiation. Therefore, its density perturbations can grow first. The resulting gravitational potential acts as an attractive potential well for ordinary matter collapsing later, speeding up the structure formation process.

## Bullet Cluster

If dark matter does not exist, then the next most likely explanation must be general relativity — the prevailing theory of gravity — is incorrect and should be modified. The Bullet Cluster, the result of a recent collision of two galaxy clusters, provides a challenge for modified gravity theories because its apparent center of mass is far displaced from the baryonic center of mass. Standard dark matter models can easily explain this observation, but modified gravity has a much harder time, especially since the observational evidence is model-independent.

## Type Ia Supernova Distance Measurements

Type Ia supernovae can be used as standard candles to measure extragalactic distances, which can in turn be used to measure how fast the universe has expanded in the past. Data indicates the universe is expanding at an accelerating rate, the cause of which is usually ascribed to dark energy. Since observations indicate the universe is almost flat, it is expected the total energy density of everything in the universe should sum to 1 ($\Omega_{tot} \approx 1$). The measured dark energy density is $\Omega_\Lambda \approx 0.690$; the observed ordinary (baryonic) matter energy density is $\Omega_b \approx 0.0482$ and the energy density of radiation is negligible. This leaves a missing $\Omega_{dm} \approx 0.258$ which nonetheless behaves like matter — dark matter.

## Sky Surveys and Baryon Acoustic Oscillations

Baryon acoustic oscillations (BAO) are fluctuations in the density of the visible baryonic matter (normal matter) of the universe on large scales. These are predicted to arise in the Lambda-CDM model due to acoustic oscillations in the photon-baryon fluid of the early universe, and can be observed in the cosmic microwave background angular power spectrum. BAOs set up a preferred length scale for baryons. As the dark matter and baryons clumped together after recombination, the effect is much weaker in the galaxy distribution in the nearby universe, but is detectable as a subtle ($\approx 1$ percent) preference for pairs of galaxies to be separated by 147 Mpc, compared to those separated by 130–160 Mpc. This feature was predicted theoretically in the 1990s and then discovered in 2005, in two large galaxy redshift surveys, the Sloan Digital Sky Survey and the 2dF Galaxy Redshift Survey. Combining the CMB observations with BAO measurements from galaxy redshift surveys provides a precise estimate of the Hubble constant and the average matter density in the Universe. The results support the Lambda-CDM model.

## Redshift-space Distortions

Large galaxy redshift surveys may be used to make a three-dimensional map of the galaxy distribution. These maps are slightly distorted because distances are estimated from observed redshifts; the redshift contains a contribution from the galaxy's so-called peculiar velocity in addition to the dominant Hubble expansion term. On average, superclusters are expanding more slowly than the cosmic mean due to their gravity, while voids are expanding faster than average. In a redshift map, galaxies in front of a supercluster have excess radial velocities towards it and have redshifts slightly higher than their distance would imply, while galaxies behind the supercluster have redshifts slightly low for their distance. This effect causes superclusters to appear squashed in the radial direction, and likewise voids are stretched. Their angular positions are unaffected. This effect is not detectable for any one structure since the true shape is not known, but can be measured by averaging over many structures. It was predicted quantitatively by Nick Kaiser in 1987, and first decisively measured in 2001 by the 2dF Galaxy Redshift Survey. Results are in agreement with the Lambda-CDM model.

## Lyman-alpha forest

In astronomical spectroscopy, the Lyman-alpha forest is the sum of the absorption lines arising from the Lyman-alpha transition of neutral hydrogen in the spectra of distant galaxies and quasars. Lyman-alpha forest observations can also constrain cosmological models. These constraints agree with those obtained from WMAP data.

## Composition of Dark Matter: Baryonic vs. Nonbaryonic

There are various hypotheses about what dark matter could consist of, as set out in the table below:

| Some dark matter hypotheses | |
|---|---|
| Light bosons | quantum chromodynamics axions |
| | axion-like particles |
| | fuzzy cold dark matter |
| neutrinos | Standard Model |
| | sterile neutrinos |
| weak scale | supersymmetry |
| | extra dimensions |
| | little Higgs |
| | effective field theory |
| | simplified models |
| other particles | Weakly interacting massive particles |
| | self-interacting dark matter |
| | superfluid vacuum theory |

| macroscopic | primordial black holes |
|---|---|
| | massive compact halo objects (MaCHOs) |
| | Macroscopic dark matter (Macros) |
| modified gravity (MOG) | modified Newtonian dynamics (MoND) |
| | Tensor–vector–scalar gravity (TeVeS) |
| | Entropic gravity |

Dark matter can refer to any substance which interacts predominantly via gravity with visible matter (e.g., stars and planets). Hence in principle it need not be composed of a new type of fundamental particle but could, at least in part, be made up of standard baryonic matter, such as protons or neutrons.[e] However, for the reasons outlined below, most scientists think the dark matter is dominated by a non-baryonic component, which is likely composed of a currently unknown fundamental particle (or similar exotic state).

## Baryonic Matter

Baryons (protons and neutrons) make up ordinary stars and planets. However, baryonic matter also encompasses less common non-primordial black holes, neutron stars, faint old white dwarfs and brown dwarfs, collectively known as massive compact halo objects (MACHOs), which can be hard to detect.

However, multiple lines of evidence suggest the majority of dark matter is not made of baryons:

- Sufficient diffuse, baryonic gas or dust would be visible when backlit by stars.

- The theory of Big Bang nucleosynthesis predicts the observed abundance of the chemical elements. If there are more baryons, then there should also be more helium, lithium and heavier elements synthesized during the Big Bang. Agreement with observed abundances requires that baryonic matter makes up between 4–5% of the universe's critical density. In contrast, large-scale structure and other observations indicate that the total matter density is about 30% of the critical density.

- Astronomical searches for gravitational microlensing in the Milky Way found at most only a small fraction of the dark matter may be in dark, compact, conventional objects (MACHOs, etc.); the excluded range of object masses is from half the Earth's mass up to 30 solar masses, which covers nearly all the plausible candidates.

- Detailed analysis of the small irregularities (anisotropies) in the cosmic microwave background. Observations by WMAP and Planck indicate that around five sixths of the total matter is in a form that interacts significantly with ordinary matter or photons only through gravitational effects.

## Non-baryonic Matter

Candidates for non-baryonic dark matter are hypothetical particles such as axions, sterile neutrinos, weakly interacting massive particles (WIMPs), gravitationally-interacting massive particles (GIMPs), supersymmetric particles, or primordial black holes. The three neutrino types already observed are indeed abundant, and dark, and matter, but because their individual masses — however uncertain they may be — are almost certainly too tiny, they can only supply a small fraction of dark matter, due to limits derived from large-scale structure and high-redshift galaxies.

Unlike baryonic matter, nonbaryonic matter did not contribute to the formation of the elements in the early universe (Big Bang nucleosynthesis) and so its presence is revealed only via its gravitational effects, or weak lensing. In addition, if the particles of which it is composed are supersymmetric, they can undergo annihilation interactions with themselves, possibly resulting in observable by-products such as gamma rays and neutrinos (indirect detection).

## Dark Matter Aggregation and Dense Dark Matter Objects

If dark matter is composed of weakly-interacting particles, an obvious question is whether it can form objects equivalent to planets, stars, or black holes. Historically, the answer has been it cannot, because of two factors:

- It lacks an efficient means to lose energy:

  Ordinary matter forms dense objects because it has numerous ways to lose energy. Losing energy would be essential for object formation, because a particle that gains energy during compaction or falling "inward" under gravity, and cannot lose it any other way, will heat up and increase velocity and momentum. Dark matter appears to lack means to lose energy, simply because it is not capable of interacting strongly in other ways except through gravity. The virial theorem suggests that such a particle would not stay bound to the gradually forming object — as the object began to form and compact, the dark matter particles within it would speed up and tend to escape.

- It lacks a range of interactions needed to form structures:

  Ordinary matter interacts in many different ways. This allows the matter to form more complex structures. For example, stars form through gravity, but the particles within them interact and can emit energy in the form of neutrinos and electromagnetic radiation through fusion when they become energetic enough. Protons and neutrons can bind via the strong interaction and then form atoms with electrons largely through electromagnetic interaction. But there is no evidence that dark matter is capable of such a wide variety of interactions, since it seems to only interact through gravity (and possibly through some means no stronger than the weak interaction, although until dark matter is better understood, this is only hopeful speculation).

In 2015–2017 the idea dense dark matter was composed of primordial black holes, made a comeback following results of gravitation wave measurements which detected the merger of intermediate mass black holes. Black holes with about 30 solar masses are not predicted to form by either stellar collapse (typically less than 15 solar masses) or by the merger of black holes in galactic centers (millions or billions of solar masses). It was proposed the intermediate mass black holes causing the detected merger formed in the hot dense early phase of the universe due to denser regions collapsing. However this was later ruled out by a survey of about a thousand supernova which detected no gravitational lensing events, although about 8 would be expected if intermediate mass primordial black holes accounted for the majority of dark matter. The possibility atom-sized primordial black holes account for a significant fraction of dark matter was ruled out by measurements of positron and electron fluxes outside the suns heliosphere by the Voyager 1 spacecraft. Tiny black holes are theorized to emit Hawking radiation. However the detected fluxes were too low and did not have the expected energy spectrum suggesting tiny primordial black holes are not widespread enough to account for dark matter. None-the-less research and theories proposing dense dark matter accounts for dark matter continue as of 2018, including approaches to dark matter cooling, and the question remains unsettled. In 2019, the lack of microlensing effects in the observation of Andromeda suggests tiny black holes do not exist.

## Classification of Dark Matter: Cold, Warm or Hot

Dark matter can be divided into *cold*, *warm*, and *hot* categories. These categories refer to velocity rather than an actual temperature, indicating how far corresponding objects moved due to random motions in the early universe, before they slowed due to cosmic expansion — this is an important distance called the free streaming length (FSL). Primordial density fluctuations smaller than this length get washed out as particles spread from overdense to underdense regions, while larger fluctuations are unaffected; therefore this length sets a minimum scale for later structure formation. The categories are set with respect to the size of a protogalaxy (an object that later evolves into a dwarf galaxy): Dark matter particles are classified as cold, warm, or hot according to their FSL; much smaller (cold), similar to (warm), or much larger (hot) than a protogalaxy.

Mixtures of the above are also possible: a theory of mixed dark matter was popular in the mid-1990s, but was rejected following the discovery of dark energy.

Cold dark matter leads to a bottom-up formation of structure with galaxies forming first and galaxy clusters at a latter stage, while hot dark matter would result in a top-down formation scenario with large matter aggregations forming early, later fragmenting into separate galaxies; the latter is excluded by high-redshift galaxy observations.

## Alternative Definitions

These categories also correspond to fluctuation spectrum effects and the interval fol-lowing the Big Bang at which each type became non-relativistic. Davis *et al.* wrote in 1985:

> "Candidate particles can be grouped into three categories on the basis of their effect on the fluctuation spectrum. If the dark matter is composed of abundant light particles which remain relativistic until shortly before recombination, then it may be termed "hot". The best candidate for hot dark matter is a neutrino. A second possibility is for the dark matter particles to interact more weakly than neutrinos, to be less abundant, and to have a mass of order 1 keV. Such parti-cles are termed "warm dark matter", because they have lower thermal velocities than massive neutrinos there are at present few candidate particles which fit this description. Gravitinos and photinos have been suggested. Any particles which became nonrelativistic very early, and so were able to diffuse a negligible distance, are termed "cold" dark matter (CDM). There are many candidates for CDM including supersymmetric particles."

> — M. Davis, G. Efstathiou, C.S. Frenk, and S.D.M. White.

Another approximate dividing line is warm dark matter became non-relativistic when the universe was approximately 1 year old and 1 millionth of its present size and in the radiation-dominated era (photons and neutrinos), with a photon temperature 2.7 million Kelvins. Standard physical cosmology gives the particle horizon size as $2\,c\,t$ (speed of light multiplied by time) in the radiation-dominated era, thus 2 light-years. A region of this size would expand to 2 million light-years today (absent structure formation). The actual FSL is approximately 5 times the above length, since it con-tinues to grow slowly as particle velocities decrease inversely with the scale factor after they become non-relativistic. In this example the FSL would correspond to 10 million light-years, or 3 megaparsecs, today, around the size containing an average large galaxy.

The 2.7 million K photon temperature gives a typical photon energy of 250 elec-tron-volts, thereby setting a typical mass scale for warm dark matter: particles much more massive than this, such as GeV–TeV mass WIMPs, would become non-relativistic much earlier than one year after the Big Bang and thus have FSLs much smaller than a protogalaxy, making them cold. Conversely, much lighter particles, such as neutrinos with masses of only a few eV, have FSLs much larger than a protogalaxy, thus qualify-ing them as hot.

## Cold Dark Matter

Cold dark matter offers the simplest explanation for most cosmological observations. It is dark matter composed of constituents with an FSL much smaller than a protogalaxy.

This is the focus for dark matter research, as hot dark matter does not seem capable of supporting galaxy or galaxy cluster formation, and most particle candidates slowed early.

The constituents of cold dark matter are unknown. Possibilities range from large objects like MACHOs (such as black holes and Preon stars) or RAMBOs (such as clusters of brown dwarfs), to new particles such as WIMPs and axions.

Studies of Big Bang nucleosynthesis and gravitational lensing convinced most cosmologists that MACHOs cannot make up more than a small fraction of dark matter. According to A. Peter: "the only *really plausible* dark-matter candidates are new particles."

The 1997 DAMA/NaI experiment and its successor DAMA/LIBRA in 2013, claimed to directly detect dark matter particles passing through the Earth, but many researchers remain skeptical, as negative results from similar experiments seem incompatible with the DAMA results.

Many supersymmetric models offer dark matter candidates in the form of the WIMPy Lightest Supersymmetric Particle (LSP). Separately, heavy sterile neutrinos exist in non-supersymmetric extensions to the standard model which explain the small neutrino mass through the seesaw mechanism.

## Warm Dark Matter

Warm dark matter comprises particles with an FSL comparable to the size of a protogalaxy. Predictions based on warm dark matter are similar to those for cold dark matter on large scales, but with less small-scale density perturbations. This reduces the predicted abundance of dwarf galaxies and may lead to lower density of dark matter in the central parts of large galaxies. Some researchers consider this a better fit to observations. A challenge for this model is the lack of particle candidates with the required mass $\approx$ 300 eV to 3000 eV.

No known particles can be categorized as warm dark matter. A postulated candidate is the sterile neutrino: A heavier, slower form of neutrino that does not interact through the weak force, unlike other neutrinos. Some modified gravity theories, such as scalar–tensor–vector gravity, require "warm" dark matter to make their equations work.

## Hot Dark Matter

Hot dark matter consists of particles whose FSL is much larger than the size of a protogalaxy. The neutrino qualifies as such particle. They were discovered independently, long before the hunt for dark matter: they were postulated in 1930, and detected in 1956. Neutrinos' mass is less than $10^{-6}$ that of an electron. Neutrinos interact with normal matter only via gravity and the weak force, making them difficult to detect (the weak force only works over a small distance, thus a neutrino triggers a weak force event only if it hits a nucleus head-on). This makes them 'weakly interacting light particles' (WILPs), as opposed to WIMPs.

The three known flavours of neutrinos are the *electron, muon,* and *tau.* Their masses are slightly different. Neutrinos oscillate among the flavours as they move. It is hard to determine an exact upper bound on the collective average mass of the three neutrinos (or for any of the three individually). For example, if the average neutrino mass were over 50 eV/c$^2$ (less than $10^{-5}$ of the mass of an electron), the universe would collapse. CMB data and other methods indicate that their average mass probably does not exceed 0.3 eV/c$^2$. Thus, observed neutrinos cannot explain dark matter.

Because galaxy-size density fluctuations get washed out by free-streaming, hot dark matter implies the first objects that can form are huge supercluster-size pancakes, which then fragment into galaxies. Deep-field observations show instead that galaxies formed first, followed by clusters and superclusters as galaxies clump together.

## Detection of Dark Matter Particles

If dark matter is made up of sub-atomic particles, then millions, possibly billions, of such particles must pass through every square centimeter of the Earth each second. Many experiments aim to test this hypothesis. Although WIMPs are popular search candidates, the Axion Dark Matter Experiment (ADMX) searches for axions. Another candidate is heavy hidden sector particles which only interact with ordinary matter via gravity.

These experiments can be divided into two classes: direct detection experiments, which search for the scattering of dark matter particles off atomic nuclei within a detector; and indirect detection, which look for the products of dark matter particle annihilations or decays.

## Direct Detection

Direct detection experiments aim to observe low-energy recoils (typically a few keVs) of nuclei induced by interactions with particles of dark matter, which (in theory) are passing through the Earth. After such a recoil the nucleus will emit energy in the form of scintillation light or phonons, as they pass through sensitive detection apparatus. To do this effectively, it is crucial to maintain a low background, and so such experiments operate deep underground to reduce the interference from cosmic rays. Examples of underground laboratories with direct detection experiments include the Stawell mine, the Soudan mine, the SNOLAB underground laboratory at Sudbury, the Gran Sasso National Laboratory, the Canfranc Underground Laboratory, the Boulby Underground Laboratory, the Deep Underground Science and Engineering Laboratory and the China Jinping Underground Laboratory.

These experiments mostly use either cryogenic or noble liquid detector technologies. Cryogenic detectors operating at temperatures below 100 mK, detect the heat produced when a particle hits an atom in a crystal absorber such as germanium. Noble liquid detectors detect scintillation produced by a particle collision in liquid xenon or argon. Cryogenic detector experiments include: CDMS, CRESST, EDELWEISS, EURECA. Noble liquid experiments include ZEPLIN, XENON, DEAP, ArDM, WARP, DarkSide, PandaX, and LUX, the Large Underground Xenon experiment. Both of these techniques focus strongly on their ability to distinguish background particles (which predominantly scatter off electrons) from dark matter particles (that scatter off nuclei). Other experiments include SIMPLE and PICASSO.

Currently there has been no well-established claim of dark matter detection from a direct detection experiment, leading instead to strong upper limits on the mass and interaction cross section with nucleons of such dark matter particles. The DAMA/NaI and more recent DAMA/LIBRA experimental collaborations have detected an annual modulation in the rate of events in their detectors, which they claim is due to dark matter. This results from the expectation that as the Earth orbits the Sun, the velocity of the detector relative to the dark matter halo will vary by a small amount. This claim is so far unconfirmed and in contradiction with negative results from other experiments such as LUX, SuperCDMS and XENON100.

A special case of direct detection experiments covers those with directional sensitivity. This is a search strategy based on the motion of the Solar System around the Galactic Center. A low-pressure time projection chamber makes it possible to access information on recoiling tracks and constrain WIMP-nucleus kinematics. WIMPs coming from the direction in which the Sun travels (approximately towards Cygnus) may then be separated from background, which should be isotropic. Directional dark matter experiments include DMTPC, DRIFT, Newage and MIMAC.

## Indirect Detection

Collage of six cluster collisions with dark matter maps. The clusters were observed in a study of how dark matter in clusters of galaxies behaves when the clusters collide.

Indirect detection experiments search for the products of the self-annihilation or decay of dark matter particles in outer space. For example, in regions of high dark matter density (e.g., the centre of our galaxy) two dark matter particles could annihilate to produce gamma rays or Standard Model particle-antiparticle pairs. Alternatively if the dark matter particle is unstable, it could decay into standard model (or other) particles.

These processes could be detected indirectly through an excess of gamma rays, antiprotons or positrons emanating from high density regions in our galaxy or others. A major difficulty inherent in such searches is that various astrophysical sources can mimic the signal expected from dark matter, and so multiple signals are likely required for a conclusive discovery.

A few of the dark matter particles passing through the Sun or Earth may scatter off atoms and lose energy. Thus dark matter may accumulate at the center of these bodies, increasing the chance of collision/annihilation. This could produce a distinctive signal in the form of high-energy neutrinos. Such a signal would be strong indirect proof of WIMP dark matter. High-energy neutrino telescopes such as AMANDA, IceCube and ANTARES are searching for this signal. The detection by LIGO in September 2015 of gravitational waves, opens the possibility of observing dark matter in a new way, particularly if it is in the form of primordial black holes.

Many experimental searches have been undertaken to look for such emission from dark matter annihilation or decay, examples of which follow. The Energetic Gamma Ray Experiment Telescope observed more gamma rays in 2008 than expected from the Milky Way, but scientists concluded this was most likely due to incorrect estimation of the telescope's sensitivity.

The Fermi Gamma-ray Space Telescope is searching for similar gamma rays. In April 2012, an analysis of previously available data from its Large Area Telescope instrument produced statistical evidence of a 130 GeV signal in the gamma radiation coming from the center of the Milky Way. WIMP annihilation was seen as the most probable explanation.

At higher energies, ground-based gamma-ray telescopes have set limits on the annihilation of dark matter in dwarf spheroidal galaxies and in clusters of galaxies.

The PAMELA experiment (launched in 2006) detected excess positrons. They could be from dark matter annihilation or from pulsars. No excess antiprotons were observed.

In 2013 results from the Alpha Magnetic Spectrometer on the International Space Station indicated excess high-energy cosmic rays which could be due to dark matter annihilation.

## Collider Searches for Dark Matter

An alternative approach to the detection of dark matter particles in nature is to produce them in a laboratory. Experiments with the Large Hadron Collider (LHC) may be able to detect dark matter particles produced in collisions of the LHC proton beams. Because a dark matter particle should have negligible interactions with normal visible matter, it may be detected indirectly as (large amounts of) missing energy and momentum that escape the detectors, provided other (non-negligible) collision products are detected. Constraints on dark matter also exist from the LEP experiment using a similar principle, but probing the interaction of dark matter particles with electrons rather than quarks. It is important to note that any discovery from collider searches must be corroborated by discoveries in the indirect or direct detection sectors to prove that the particle discovered is, in fact, dark matter.

## Alternative Hypotheses

Because dark matter remains to be conclusively identified, many other hypotheses have emerged aiming to explain the observational phenomena that dark matter was conceived to explain. The most common method is to modify general relativity. General relativity is well-tested on solar system scales, but its validity on galactic or cosmological scales has not been well proven. A suitable modification to general relativity can conceivably eliminate the need for dark matter. The best-known theories of this class are MOND and its relativistic generalization tensor-vector-scalar gravity (TeVeS), f(R) gravity, negative mass dark fluid, and entropic gravity. Alternative theories abound.

A problem with alternative hypotheses is observational evidence for dark matter comes from so many independent approaches.Explaining any individual observation is possible but explaining all of them is very difficult. Nonetheless, there have been some scattered successes for alternative hypotheses, such as a 2016 test of gravitational lensing in entropic gravity.

The prevailing opinion among most astrophysicists is while modifications to general relativity can conceivably explain part of the observational evidence, there is probably enough data to conclude there must be some form of dark matter.

## Dark Energy

Dark Energy is a hypothetical form of energy that exerts a negative, repulsive pressure, behaving like the opposite of gravity. It has been hypothesised to account for the observational properties of distant type Ia supernovae, which show the universe going through an accelerated period of expansion. Like Dark Matter, Dark Energy is not directly observed, but rather inferred from observations of gravitational interactions between astronomical objects.

Dark Energy makes up 72% of the total mass-energy density of the universe. The other dominant contributor is Dark Matter, and a small amount is due to atoms or baryonic matter.

In 1998 two teams of astronomers announced that distant, $z \sim 1$ type Ia supernovae were slightly too faint than model predictions of an expanding (yet slowing) universe. To be fainter, the supernovae must be farther away and this requires that the expansion of the Universe was slower in the past. Both teams agreed that the universe is going through a phase of accelerated expansion. Dark Energy was invoked to drive this acceleration.

In the early part of the 20th century Albert Einstein had invoked a 'cosmological constant', (usually symbolized by the Greek letter lambda, $\Lambda$). It was a vacuum energy of empty space, which kept the universe (predicted by his field equations of the General Theory of Relativity) static, rather than contracting or expanding. It provided a way of balancing the gravitational contraction caused by matter. Once the universe was observed to be expanding Einstein hastily removed his cosmological constant. However if dark energy is described by something similar to Einstein's cosmological constant it doesn't just balance gravity to keep a static universe but has negative pressure to cause the expansion to accelerate.

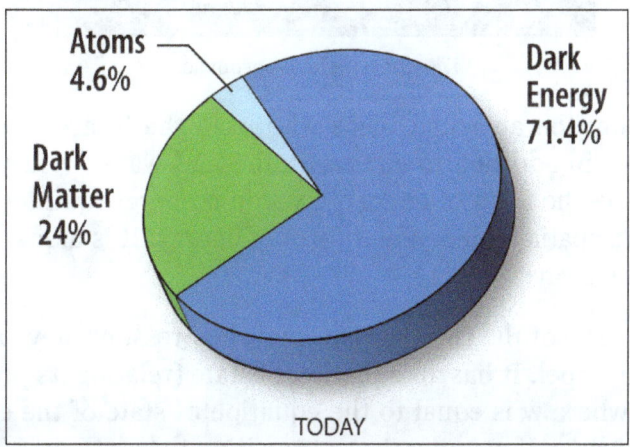

Schematic representation of the total mass-energy density in the universe.

Other types of dark energy have been proposed, including a cosmic field associated with inflation and a different, low-energy field dubbed "quintessence".

It is thought that the very early universe also went through a period of rapid expansion, called inflation. Inflation, occurring about $10^{-36}$ seconds after the Big Bang, acted to smooth out the universe and make it geometrically flat. If the density of the universe exactly equals the critical density, then the geometry of the universe is flat like a sheet of paper. For a matter dominated Universe the critical density (equivalent to about 6 protons per m³) sits exactly between the density required for a heavy universe that will eventually collapse, and a density required for a light universe that will expand forever. When astronomers measure the amount of matter and energy in the universe today they only come up with about ~30% of what is needed to make the universe flat. The addition of Dark Energy to the mass-energy budget makes the universe flat. The simplest version of inflation, predicts that the density of the universe is very close to the critical density.

The WMAP spacecraft has measured the geometry of the universe. If the universe was flat, the brightest cosmic microwave background fluctuations (or "spots") would be about 1 degree across. WMAP has confirmed this spot size with very high accuracy. We now know that the universe is flat with only a 2% margin of error.

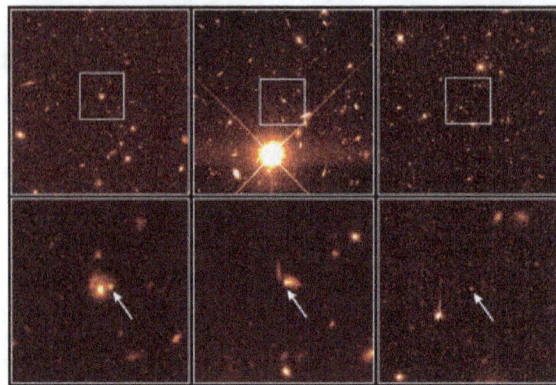

Distant type Ia supernovae.

Quintessence is from the ancient Greeks who used the term to describe a mysterious 'fifth element' – in addition to air, earth, fire and water. Whereas the cosmological constant is a specific form of energy, a vacuum energy, quintessence is dynamic, time-evolving and a spatially dependent form of energy. It is a quantum field with kinetic and potential energy.

Depending on the ratio of the two energies and the pressure they exert, quintessence can either attract or repel. It has an equation of state (relating its pressure p and density ρ) of $p = w\rho$, where w is equal to the equation of state of the energy component dominating the universe. If w undergoes a transition to less than -1/3 this initiates accelerated expansion. By contrast, a cosmological constant is static, with a fixed energy density and $w = -1$.

There are a number of ongoing programs aimed at discovering more about Dark Energy. One such study involves the measurement of Baryonic Acoustic Oscillations (BAO).

Alternatives to Dark Energy have been proposed. Some scientists have proposed that our Galaxy sits inside a region of low density caused by the passage of a density wave. The Big Bang may have created this large-scale wave in space-time. As this primordial wave moved through the universe, it left behind a low-density ripple several tens of millions of light-years across, in which the Galaxy now resides. Whilst possible, this difference in the properties of space-time would violate the Copernican principle which states that the universe, on large scales is homogenous.

## Accretion

In astrophysics, accretion is the accumulation of particles into a massive object by gravitationally attracting more matter, typically gaseous matter, in an accretion disk. Most astronomical objects, such as galaxies, stars, and planets, are formed by accretion processes.

ALMA image of HL Tauri, a protoplanetary disk.

The accretion model that Earth and the other terrestrial planets formed from meteoric material was proposed in 1944 by Otto Schmidt, followed by the *protoplanet theory* of William McCrea (1960) and finally the *capture theory* of Michael Woolfson. In 1978, Andrew Prentice resurrected the initial Laplacian ideas about planet formation and developed the *modern Laplacian theory*. None of these models proved completely successful, and many of the proposed theories were descriptive.

The 1944 accretion model by Otto Schmidt was further developed in a quantitative way in 1969 by Viktor Safronov. He calculated, in detail, the different stages of terrestrial planet formation. Since then, the model has been further developed using intensive numerical simulations to study planetesimal accumulation. It is now accepted that stars form by the gravitational collapse of interstellar gas. Prior to collapse, this gas is mostly in the form of molecular clouds, such as the Orion Nebula. As the cloud collapses, losing potential energy, it heats up, gaining kinetic energy, and the conservation of angular momentum ensures that the cloud forms a flatted disk—the accretion disk.

## Accretion of Galaxies

A few hundred thousand years after the Big Bang, the Universe cooled to the point where atoms could form. As the Universe continued to expand and cool, the atoms lost enough kinetic energy, and dark matter coalesced sufficiently, to form protogalaxies. As further accretion occurred, galaxies formed. Indirect evidence is widespread. Galaxies grow through mergers and smooth gas accretion. Accretion also occurs inside galaxies, forming stars.

## Accretion of Stars

Stars are thought to form inside giant clouds of cold molecular hydrogen—giant molecular clouds of roughly 300,000 $M_\odot$ and 65 light-years (20 pc) in diameter. Over millions of years, giant molecular clouds are prone to collapse and fragmentation. These fragments then form small, dense cores, which in turn collapse into stars. The cores range in mass from a fraction to several times that of the Sun and are called protostellar (protosolar) nebulae. They possess diameters of 2,000–20,000 astronomical units (0.01–0.1 pc) and a particle number density of roughly 10,000 to 100,000/cm³ (160,000 to 1,600,000/cu in). Compare it with the particle number density of the air at the sea level—$2.8 \times 10^{19}$/cm³ ($4.6 \times 10^{20}$/cu in).

The visible-light (left) and infrared (right) views of the Trifid Nebula, a giant star-forming cloud of gas and dust located 5,400 light-years (1,700 pc) away in the constellation Sagittarius.

The initial collapse of a solar-mass protostellar nebula takes around 100,000 years. Every nebula begins with a certain amount of angular momentum. Gas in the central part of the nebula, with relatively low angular momentum, undergoes fast compression and forms a hot hydrostatic (non-contracting) core containing a small fraction of the mass of the original nebula. This core forms the seed of what will become a star. As the collapse continues, conservation of angular momentum dictates that the rotation of the infalling envelope accelerates, which eventually forms a disk.

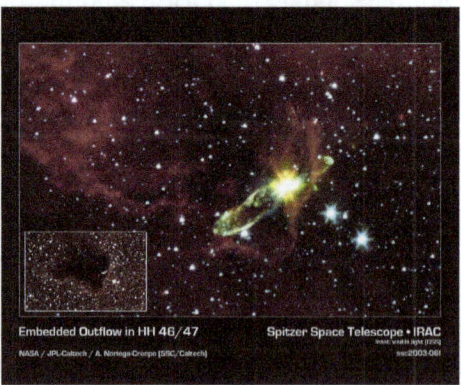

Infrared image of the molecular outflow from an otherwise hidden newborn star HH 46/47.

As the infall of material from the disk continues, the envelope eventually becomes thin and transparent and the young stellar object (YSO) becomes observable, initially in far-infrared light and later in the visible. Around this time the protostar begins to fuse deuterium. If the protostar is sufficiently massive (above 80 $M_J$), hydrogen fusion follows. Otherwise, if its mass is too low, the object becomes a brown dwarf. This birth of a new star occurs approximately 100,000 years after the collapse begins. Objects at this stage are known as Class I protostars, which are also called young T Tauri stars, evolved protostars, or young stellar objects. By this time, the forming star has already accreted much of its mass; the total mass of the disk and remaining envelope does not exceed 10–20% of the mass of the central YSO.

When the lower-mass star in a binary system enters an expansion phase, its outer atmosphere may fall onto the compact star, forming an accretion disk.

At the next stage, the envelope completely disappears, having been gathered up by the disk, and the protostar becomes a classical T Tauri star. The latter have accretion disks and continue to accrete hot gas, which manifests itself by strong emission lines in their spectrum. The former do not possess accretion disks. Classical T Tauri stars evolve into weakly lined T Tauri stars. This happens after about 1 million years. The mass of the disk around a classical T Tauri star is about 1–3% of the stellar mass, and it is accreted at a rate of $10^{-7}$ to $10^{-9}$ $M_\odot$ per year. A pair of bipolar jets is usually present as well. The accretion explains all peculiar properties of classical T Tauri stars: strong flux in the emission lines (up to 100% of the intrinsic luminosity of the star), magnetic activity, photometric variability and jets. The emission lines actually form as the accreted gas hits the "surface" of the star, which happens around its magnetic poles. The jets are byproducts of accretion: they carry away excessive angular momentum. The classical T Tauri stage lasts about 10 million years. The disk eventually disappears due to accretion onto the central star, planet formation, ejection by jets, and photoevaporation by ultraviolet radiation from the central star and nearby stars. As a result, the young star becomes a weakly lined T Tauri star, which, over hundreds of millions of years, evolves into an ordinary Sun-like star, dependent on its initial mass.

## Accretion of Planets

Self-accretion of cosmic dust accelerates the growth of the particles into boulder-sized planetesimals. The more massive planetesimals accrete some smaller ones, while others shatter in collisions. Accretion disks are common around smaller stars, or stellar remnants in a close binary, or black holes surrounded by material, such as those at the centers of galaxies. Some dynamics in the disk, such as dynamical friction, are necessary to allow orbiting gas to lose angular momentum and fall onto the central massive object. Occasionally, this can result in stellar surface fusion.

Artist's impression of a protoplanetary disk showing a young star at its center

In the formation of terrestrial planets or planetary cores, several stages can be considered. First, when gas and dust grains collide, they agglomerate by microphysical processes like van der Waals forces and electromagnetic forces, forming micrometer-sized particles; during this stage, accumulation mechanisms are largely non-gravitational in nature. However, planetesimal formation in the centimeter-to-meter range is not well understood, and no convincing explanation is offered as to why such grains would accumulate rather than simply rebound. In particular, it is still not clear how these objects grow to become 0.1–1 km (0.06–0.6 mi) sized planetesimals; this problem is known as the "meter size barrier": As dust particles grow by coagulation, they acquire increasingly large relative velocities with respect to other particles in their vicinity, as well as a systematic inward drift velocity, that leads to destructive collisions, and thereby limit the growth of the aggregates to some maximum size. Ward (1996) suggests that when slow moving grains collide, the very low, yet non-zero, gravity of colliding grains impedes their escape. It is also thought that grain fragmentation plays an important role replenishing small grains and keeping the disk thick, but also in maintaining a relatively high abundance of solids of all sizes.

A number of mechanisms have been proposed for crossing the 'meter-sized' barrier. Local concentrations of pebbles may form, which then gravitationally collapse into planetesimals the size of large asteroids. These concentrations can occur passively due to the structure of the gas disk, for example, between eddies, at pressure bumps, at the edge of a gap created by a giant planet, or at the boundaries of turbulent regions of the disk. Or, the particles may take an active role in their concentration via a feedback mechanism referred to as a streaming instability. In a streaming instability the interaction between the solids and the gas in the protoplanetary disk results in the growth of local concentrations, as new particles accumulate in the wake of small concentrations, causing them to grow into massive filaments. Alternatively, if the grains that form due to the agglomeration of dust are highly porous their growth may continue until they become large enough to collapse due to their own gravity. The low density of these objects allows them to remain strongly coupled with the gas, thereby avoiding high velocity collisions which could result in their erosion or fragmentation.

Grains eventually stick together to form mountain-size (or larger) bodies called planetesimals. Collisions and gravitational interactions between planetesimals combine to produce Moon-size planetary embryos (protoplanets) over roughly 0.1–1 million years. Finally, the planetary embryos collide to form planets over 10–100 million years. The planetesimals are massive enough that mutual gravitational interactions are significant enough to be taken into account when computing their evolution. Growth is aided by orbital decay of smaller bodies due to gas drag, which prevents them from being stranded between orbits of the embryos. Further collisions and accumulation lead to terrestrial planets or the core of giant planets.

If the planetesimals formed via the gravitational collapse of local concentrations of pebbles their growth into planetary embryos and the cores of giant planets is dominated by the further accretions of pebbles. Pebble accretion is aided by the gas drag felt by objects as they accelerate toward a massive body. Gas drag slows the pebbles below the escape velocity of the massive body causing them to spiral toward and to be accreted by it. Pebble accretion may accelerate the formation of planets by a factor of 1000 compared to the accretion of planetesimals, allowing giant planets to form before the dissipation of the gas disk. Yet, core growth via pebble accretion appears incompatible with the final masses and compositions of Uranus and Neptune.

The formation of terrestrial planets differs from that of giant gas planets, also called Jovian planets. The particles that make up the terrestrial planets are made from metal and rock that condense in the inner Solar System. However, Jovian planets begin as large, icy planetesimals, which then capture hydrogen and helium gas from the solar nebula. Differentiation between these two classes of planetesimals arise due to the frost line of the solar nebula.

## Accretion of Asteroids

Chondrules in a chondrite meteorite. A millimeter scale is shown.

Meteorites contain a record of accretion and impacts during all stages of asteroid origin and evolution; however, the mechanism of asteroid accretion and growth is not well understood. Evidence suggests the main growth of asteroids can result from gas-assisted accretion of chondrules, which are millimeter-sized spherules that form as molten (or partially molten) droplets in space before being accreted to their parent asteroids. In the inner Solar System, chondrules appear to have been crucial for initiating accretion. The tiny mass of asteroids may be partly due to inefficient chondrule formation beyond 2 AU, or less-efficient delivery of chondrules from near the protostar. Also, impacts controlled the formation and destruction of asteroids, and are thought to be a major factor in their geological evolution.

Chondrules, metal grains, and other components likely formed in the solar nebula. These accreted together to form parent asteroids. Some of these bodies subsequently melted, forming metallic cores and olivine-rich mantles; others were aqueously altered. After the asteroids had cooled, they were eroded by impacts for 4.5 billion years, or disrupted.

For accretion to occur, impact velocities must be less than about twice the escape velocity, which is about 140 m/s (460 ft/s) for a 100 km (60 mi) radius asteroid. Simple models for accretion in the asteroid belt generally assume micrometer-sized dust grains sticking together and settling to the midplane of the nebula to form a dense layer of dust, which, because of gravitational forces, was converted into a disk of kilometer-sized planetesimals. But, several arguments suggest that asteroids may not have accreted this way.

## Accretion of Comets

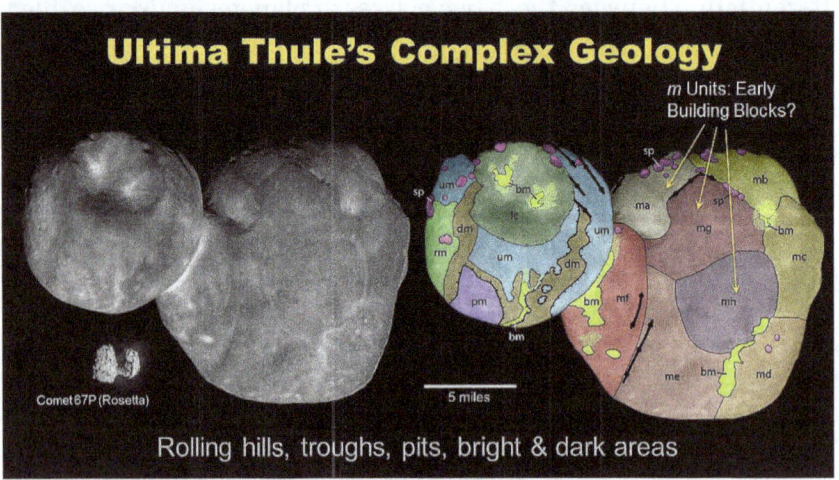

The geology of 2014 $MU_{69}$. Notable surface features are highlighted.

Comets, or their precursors, formed in the outer Solar System, possibly millions of years before planet formation. How and when comets formed is debated, with distinct implications for Solar System formation, dynamics, and geology. Three-dimensional computer simulations indicate the major structural features observed on cometary nuclei can be explained by pairwise low velocity accretion of weak cometesimals. The currently favored formation mechanism is that of the nebular hypothesis, which states that comets are probably a remnant of the original planetesimal "building blocks" from which the planets grew.

Astronomers think that comets originate in both the Oort cloud and the scattered disk. The scattered disk was created when Neptune migrated outward into the proto-Kuiper belt, which at the time was much closer to the Sun, and left in its wake a population of dynamically stable objects that could never be affected by its orbit (the Kuiper belt proper), and a population whose perihelia are close enough that Neptune can still disturb them as it travels around the Sun (the scattered disk). Because the scattered disk is dynamically active and the Kuiper belt relatively dynamically stable, the scattered disk is now seen as the most likely point of origin for periodic comets. The classic Oort cloud theory states that the Oort cloud, a sphere measuring about 50,000 AU (0.24 pc) in radius, formed at the same time as the solar nebula and occasionally releases comets into the inner Solar System as a giant planet or star passes nearby and causes gravitational disruptions. Examples of such comet clouds may already have been seen in the Helix Nebula.

The *Rosetta* mission to comet 67P/Churyumov–Gerasimenko determined in 2015 that when Sun's heat penetrates the surface, it triggers evaporation (sublimation) of buried ice. While some of the resulting water vapour may escape from the nucleus, 80% of it recondenses in layers beneath the surface. This observation implies that the thin ice-rich layers exposed close to the surface may be a consequence of cometary activity and evolution, and that global layering does not necessarily occur early in the comet's formation history. While most scientists thought that all the evidence indicated that the structure of nuclei of comets is processed rubble piles of smaller ice planetesimals of a previous generation, the *Rosetta* mission dispelled the idea that comets are "rubble piles" of disparate material.

## Optical Depth

Optical depth in astrophysics refers to a specific level of transparency. Optical depth and actual depth, $\tau$ and $z$ respectively, can vary widely depending on the absorptivity of the astrophysical environment. Indeed, $\tau$ is able to show the relationship between these two quantities and can lead to a greater understanding of the structure inside a star.

Optical depth is a measure of the extinction coefficient or absorptivity up to a specific 'depth' of a star's makeup.

$$\tau \equiv \int_0^z \alpha \, dz = \sigma N.$$

The assumption here is that either the extinction coefficient $\alpha$ or the column number density $N$ is known. These can generally be calculated from other equations if a fair amount of information is known about the chemical makeup of the star. From the definition, it is also clear that large optical depths correspond to higher rate of obscuration. Optical depth can therefore be thought of as the opacity of a medium.

The extinction coefficient $\alpha$ can be calculated using the transfer equation. In most astrophysical problems, this is exceptionally difficult to solve since solving the corresponding equations requires the incident radiation as well as the radiation leaving the star. These values are usually theoretical.

In some cases the Beer-Lambert Law can be useful in finding $\alpha$.

$$\alpha = e^{\frac{4\pi\kappa}{\lambda_0}},$$

where $\kappa$ is the refractive index, and $\lambda_0$ is the wavelength of the incident light before being absorbed or scattered. It is important to note that the Beer-Lambert Law is only appropriate when the absorption occurs at a specific wavelength, $\lambda_0$. For a gray atmosphere, for instance, it is most appropriate to use the Eddington Approximation.

Therefore, $\tau$ is simply a constant that depends on the physical distance from the outside of a star. To find $\tau$ at a particular depth $z'$, the above equation may be used with $\alpha$ and integration from $z = 0$ to $z = z'$.

## The Eddington Approximation and the Depth of the Photosphere

Since it is difficult to define where the photosphere of a star ends and the chromosphere begins, astrophysicists usually rely on the Eddington Approximation to derive the formal definition of $\tau = 2/3$.

Devised by Sir Arthur Eddington the approximation takes into account the fact that $H^-$ produces a "gray" absorption in the atmosphere of a star, that is, it is independent of any specific wavelength and absorbs along the entire electromagnetic spectrum. In that case,

$$T^4 = \frac{3}{4}T_e^4\left(\tau + \frac{2}{3}\right),$$

where $T_e$ is the effective temperature at that depth and $\tau$ is the optical depth.

This illustrates not only that the observable temperature and actual temperature at a certain physical depth of a star vary, but that the optical depth plays a crucial role in understanding the stellar structure. It also serves to demonstrate that the depth of the photosphere of a star is highly dependent upon the absorptivity of its environment. The photosphere extends down to a point where is about 2/3, which corresponds to a state where a photon would experience, in general, less than 1 scattering before leaving the star.

The above equation can be rewritten in terms of $\alpha$ in the following way:

$$T^4 = \frac{3}{4}T_e^4\left(\int_0^z (\alpha)dz + \frac{2}{3}\right)$$

Which is useful, for example, when $\tau$ is not known but $\alpha$ is.

## References

- Cords, scholarx, spider: seds.org, Retrieved 4 February 2019

- Leverington, David (2003). Babylon to Voyager and beyond: a history of planetary astronomy. Cambridge University Press. P. 126. ISBN 978-0-521-80840-8

- IAU WG on NSFA Current Best Estimates, archived from the original on 8 December 2009, retrieved 25 September 2009

- What-is-the-big-bang-theory: thoughtco.com, 09 June 2019

- Fienga, A.; et al. (2011), "The INPOP10a planetary ephemeris and its applications in fundamental physics", Celestial Mechanics and Dynamical Astronomy, 111 (3): 363, arXiv:1108.5546, Bibcode:2011CeMDA.111..363F, doi:10.1007/s10569-011-9377-8

- Holton, Gerald James; Brush, Stephen G. (2001). Physics, the human adventure: from Copernicus to Einstein and beyond (3rd ed.). Rutgers University Press. P. 137. ISBN 978-0-8135-2908-0

- Dark-Energy, cosmos: swin.edu.au, Retrieved 1 May, 2019

# Astronomical Objects

**4**

- **Celestial Bodies**
- **Star**
- **Galaxy**
- **Nebula**
- **Interstellar Medium**

The naturally occurring physical entity or structure that exists in the observable universe is known as an astronomical object. It includes celestial objects, stars, galaxies, nebulas, star clusters, etc. This chapter has been carefully written to provide an easy understanding of the various aspects of these astronomical objects.

## Celestial Bodies

Celestial bodies or heavenly bodies are objects in space such as the sun, moon, planets and stars. They form a part of the vast universe we live in and are usually very far from us. The glorious night sky is dotted with such objects and when we observe them using a telescope they reveal fascinating worlds of their own. Because they are so far away, we cannot see all of them using the naked eye and we rely upon telescopes to study them.

Hence, we can define heavenly bodies as,

A planet, moon, star or other natural objects in the space.

### Classification of Celestial Bodies

### Stars

Stars are giant balls of hot gases that can produce their own light. Stars give out energy by converting Hydrogen gas into Helium in their cores. Stars are gigantic in size and have immense gravitational attraction. The sun is a medium-sized star that gives us energy and makes life possible on earth.

## Planets

Planets are large (almost) spherical objects that revolve around the sun. Planets move in fixed orbits around the sun. There are 8 planets in our solar system. Planets may be made of rocks, metals and gases like hydrogen, nitrogen and methane. The earth is also a planet and is the only known place in the universe which supports life. Planets that revolve around other stars are called exoplanets.

## Satellites

Satellites are objects that revolve around planets. They form the essential part of the celestial bodies. These may be of natural origin or sent by humans. The moon is a natural satellite of the earth and revolves around it because it is bound by the Earth's gravitational pull. Man has also placed artificial or man-made satellites around the earth and other planets to study them and for communication purposes.

## Comets

These are small chunks of ice and rock that come from the outer edge of the solar system. When its orbit brings it closer to the sun, the ice on them vaporizes creating a beautiful tail behind them.

## Asteroids

These are small irregularly shaped rocks made up of metal or minerals that orbit the sun. Most of them are found between Mars and Jupiter in an area known as the asteroid belt.

## Meteors and Meteorites

These are objects from space that enter our atmosphere as they are pulled by the earth's gravity. Meteors usually are small and burn up in the atmosphere as they enter the earth. This creates streaks in the sky as though a star has fallen. They are commonly called *shooting stars*. If a meteor is large enough it can reach the ground and create a crater. Such objects are called meteorites.

## Galaxies

Galaxies are large groups of stars held together by gravity. The sun and the solar system is part of a galaxy known as the Milky Way. Other galaxies are usually so far away that they look like stars in the night sky. The Andromeda galaxy and the Large Magellanic Cloud are galaxies that can be seen with the naked eye on a clear night.

(1) Stars    (2) Planets    (3) Satellites

(4) Comets    (5) Asteroids    (6) Meteors and meteorites

(7) Galaxies

## Star

Star is any massive self-luminous celestial body of gas that shines by radiation derived from its internal energy sources. Of the tens of billions of trillions of stars composing the observable universe, only a very small percentage are visible to the naked eye. Many stars occur in pairs, multiple systems, or star clusters. The members of such stellar groups are physically related through common origin and are bound by mutual gravitational attraction. Somewhat related to star clusters are stellar associations, which consist of loose groups of physically similar stars that have insufficient mass as a group to remain together as an organization.

### The Sun as a Point of Comparison

### Variations in Stellar Size

With regard to mass, size, and intrinsic brightness, the Sun is a typical star. Its approximate mass is $2 \times 10^{30}$ kg (about 330,000 Earth masses), its approximate radius 700,000 km (430,000 miles), and its approximate luminosity $4 \times 10^{33}$ ergs per second (or equivalently $4 \times 10^{23}$ kilowatts of power). Other stars often have their respective quantities measured in terms of those of the Sun.

Many stars vary in the amount of light they radiate. Stars such as Altair, Alpha Centauri A and B, and Procyon A are called dwarf stars; their dimensions are roughly comparable to those of the Sun. Sirius A and Vega, though much brighter, also are dwarf stars; their higher temperatures yield a larger rate of emission per unit area. Aldebaran A, Arcturus, and Capella A are examples of giant stars, whose dimensions are much larger than those of the Sun. Observations with an interferometer (an instrument that measures the angle subtended by the diameter of a star at the observer's position), combined with parallax measurements, give sizes of 12 and 22 solar radii for Arcturus and Aldebaran A. Betelgeuse and Antares A are examples of supergiant stars. The latter has a radius some 300 times that of the Sun, whereas the variable star Betelgeuse oscillates between roughly 300 and 600 solar radii. Several of the stellar class of white dwarf stars, which have low luminosities and high densities, also are among the brightest stars. Sirius B is a prime example, having a radius one-thousandth that of the Sun, which is comparable to the size of Earth. Also among the brightest stars are Rigel A, a young supergiant in the constellation Orion, and Canopus, a bright beacon in the Southern Hemisphere often used for spacecraft navigation.

## Stellar Activity and Mass Loss

The Sun's activity is apparently not unique. It has been found that stars of many types are active and have stellar winds analogous to the solar wind. The importance and ubiquity of strong stellar winds became apparent only through advances in spaceborne ultraviolet and X-ray astronomy as well as in radio and infrared surface-based astronomy.

X-ray observations that were made during the early 1980s yielded some rather unexpected findings. They revealed that nearly all types of stars are surrounded by coronas having temperatures of one million kelvins (K) or more. Furthermore, all stars seemingly display active regions, including spots, flares, and prominences much like those of the Sun.

Some stars exhibit starspots so large that an entire face of the star is relatively dark, while others display flare activity thousands of times more intense than that on the Sun.

The highly luminous hot, blue stars have by far the strongest stellar winds. Observations of their ultraviolet spectra with telescopes on sounding rockets and spacecraft have shown that their wind speeds often reach 3,000 km (roughly 2,000 miles) per second, while losing mass at rates up to a billion times that of the solar wind. The corresponding mass-loss rates approach and sometimes exceed one hundred-thousandth of a solar mass per year, which means that one entire solar mass (perhaps a tenth of the total mass of the star) is carried away into space in a relatively short span of 100,000 years. Accordingly, the most luminous stars are thought to lose substantial fractions of their mass during their lifetimes, which are calculated to be only a few million years.

Ultraviolet observations have proved that to produce such great winds the pressure of hot gases in a corona, which drives the solar wind, is not enough. Instead, the winds of the hot stars must be driven directly by the pressure of the energetic ultraviolet radiation emitted by these stars. Aside from the simple realization that copious quantities of ultraviolet radiation flow from such hot stars, the details of the process are not well understood. Whatever is going on, it is surely complex, for the ultraviolet spectra of the stars tend to vary with time, implying that the wind is not steady. In an effort to understand better the variations in the rate of flow, theorists are investigating possible kinds of instabilities that might be peculiar to luminous hot stars.

Observations made with radio and infrared telescopes as well as with optical instruments prove that luminous cool stars also have winds whose total mass-flow rates are comparable to those of the luminous hot stars, though their velocities are much lower—about 30 km (20 miles) per second. Because luminous red stars are inherently cool objects (having a surface temperature of about 3,000 K, or half that of the Sun), they emit very little detectable ultraviolet or X-ray radiation; thus, the mechanism driving the winds must differ from that in luminous hot stars. Winds from luminous cool stars, unlike those from hot stars, are rich in dust grains and molecules. Since nearly all stars more massive than the Sun eventually evolve into such cool stars, their winds, pouring into space from vast numbers of stars, provide a major source of new gas and dust in interstellar space, thereby furnishing a vital link in the cycle of star formation and galactic evolution. As in the case of the hot stars, the specific mechanism that drives the winds of the cool stars is not understood; at this time, investigators can only surmise that gas turbulence, magnetic fields, or both in the atmospheres of these stars are somehow responsible.

Strong winds also are found to be associated with objects called protostars, which are huge gas balls that have not yet become full-fledged stars in which energy is provided by nuclear reactions. Radio and infrared observations of deuterium (heavy hydrogen) and carbon monoxide (CO) molecules in the Orion Nebula have revealed clouds of gas expanding outward at velocities approaching 100 km (60 miles) per second. Furthermore, high-resolution, very-long-baseline interferometry observations have disclosed expanding knots of natural maser (coherent microwave) emission of water vapour near the star-forming regions in Orion, thus linking the strong winds to the protostars themselves. The specific causes of these winds remain unknown, but if they generally accompany star formation, astronomers will have to consider the implications for the early solar system. After all, the Sun was presumably once a protostar too.

## Distances to the Stars

### Determining Stellar Distances

Distances to stars were first determined by the technique of trigonometric parallax, a method still used for nearby stars. When the position of a nearby star is measured from two points on opposite sides of Earth's orbit (i.e., six months apart), a small angular (artificial) displacement is observed relative to a background of very remote (essentially fixed) stars. Using the radius of Earth's orbit as the baseline, the distance of the star can be found from the parallactic angle, p. If p = 1″ (one second of arc), the distance of the star is 206,265 times Earth's distance from the Sun—namely, 3.26 light-years. This unit of distance is termed the parsec, defined as the distance of an object whose parallax equals one arc second. Therefore, one parsec equals 3.26 light-years. Since parallax is inversely proportional to distance, a star at 10 parsecs would have a parallax of 0.1″. The nearest star to Earth, Proxima Centauri (a member of the triple system of Alpha Centauri), has a parallax of 0.76813″, meaning that its distance is 1/0.76813, or 1.302, parsecs, which equals 4.24 light-years. The parallax of Barnard's star, the next closest after the Alpha Centauri system, is 0.54831″, so that its distance is nearly 6 light-years. Errors of such parallaxes are now typically 0.001″. Thus, measurements of trigonometric parallaxes are useful for only the nearby stars within a few thousand light-years. In fact, of the approximately 100 billion stars in the Milky Way Galaxy (also simply called the Galaxy), the Hipparcos satellite has measured only about 100,000 to an accuracy of 0.001″. For more distant stars, indirect methods are used; most of them depend on comparing the intrinsic brightness of a star (found, for example, from its spectrum or other observable property) with its apparent brightness.

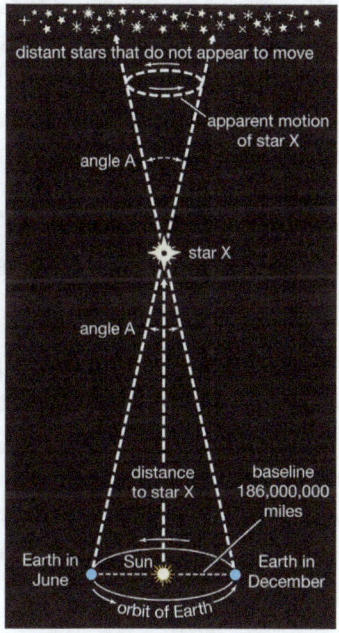

Calculating stellar distances.

## Nearest Stars

Only three stars, Alpha Centauri, Procyon, and Sirius, are both among the 20 nearest and among the 20 brightest stars. Ironically, most of the relatively nearby stars are dimmer than the Sun and are invisible without the aid of a telescope. By contrast, some of the well-known bright stars outlining the constellations have parallaxes as small as the limiting value of 0.001″ and are therefore well beyond several hundred light-years' distance from the Sun. The most luminous stars can be seen at great distances, whereas the intrinsically faint stars can be observed only if they are relatively close to Earth.

Sirius A and B (lower left) photographed by the Hubble Space Telescope.

Although the lists of the brightest and the nearest stars pertain to only a very small number of stars, they nonetheless serve to illustrate some important points. The stars listed fall roughly into three categories: (1) giant stars and supergiant stars having sizes of tens or even hundreds of solar radii and extremely low average densities—in fact, several orders of magnitude less than that of water (one gram per cubic centimetre); (2) dwarf stars having sizes ranging from 0.1 to 5 solar radii and masses from 0.1 to about 10 solar masses; and (3) white dwarf stars having masses comparable to that of the Sun but dimensions appropriate to planets, meaning that their average densities are hundreds of thousands of times greater than that of water.

These rough groupings of stars correspond to stages in their life histories. The second category is identified with what is called the main sequence and includes stars that emit energy mainly by converting hydrogen into helium in their cores. The first category comprises stars that have exhausted the hydrogen in their cores and are burning hydrogen within a shell surrounding the core. The white dwarfs represent the final stage in the life of a typical star, when most available sources of energy have been exhausted and the star has become relatively dim.

The large number of binary stars and even multiple systems is notable. These star systems exhibit scales comparable in size to that of the solar system. Some, and perhaps many, of the nearby single stars have invisible (or very dim) companions detectable by their gravitational effects on the primary star; this orbital motion of the unseen member causes the visible star to "wobble" in its motion through space. Some of the invisible companions have been found to have masses on the order of 0.001 solar mass or less, which is in the range of planetary rather than stellar dimensions. Current observations suggest that they are genuine planets, though some are merely extremely dim stars (sometimes called brown dwarfs). Nonetheless, a reasonable inference that can be drawn from these data is that double stars and planetary systems are formed by similar evolutionary processes.

## Stellar Positions

### Basic Measurements

Accurate observations of stellar positions are essential to many problems of astrono-my. Positions of the brighter stars can be measured very accurately in the equatorial system (the coordinates of which are called right ascension [α, or RA] and declina-tion [δ, or DEC] and are given for some epoch—for example, 1950.0 or, currently, 2000.0). Positions of fainter stars are measured by using electronic imaging devices (e.g., a charge-coupled device, or CCD) with respect to the brighter stars, and, finally, the entire group is referred to the positions of known external galaxies. These distant galaxies are far enough away to define an essentially fixed, or immovable, system, whereas in the Milky Way the positions of both bright and faint stars are affected over relatively short periods of time by galactic rotation and by their own motions through the Galaxy.

### Stellar Motions

Accurate measurements of position make it possible to determine the movement of a star across the line of sight (i.e., perpendicular to the observer)—its proper motion. The amount of proper motion, denoted by μ (in arc seconds per year), divided by the parallax of the star and multiplied by a factor of 4.74 equals the tangential velocity, $V_T$, in kilometres per second in the plane of the celestial sphere.

The motion along the line of sight (i.e., toward the observer), called radial velocity, is ob-tained directly from spectroscopic observations. If λ is the wavelength of a characteristic spectral line of some atom or ion present in the star and $\lambda_L$ is the wavelength of the same line measured in the laboratory, then the difference Δλ, or $\lambda - \lambda_L$, divided by $\lambda_L$ equals the radial velocity, $V_R$, divided by the velocity of light, c—namely, $\Delta\lambda/\lambda_L = V_R/c$. Shifts of a spectral line toward the red end of the electromagnetic spectrum (i.e., positive $V_R$) indicate recession, and those toward the blue end (negative $V_R$) indicate approach. If the parallax is known, measurements of μ and $V_R$ enable a determination of the space motion of the star. Normally, radial velocities are corrected for Earth's rotation and for its motion around the Sun, so that they refer to the line-of-sight motion of the star with respect to the Sun.

Consider a pertinent example. The proper motion of Alpha Centauri is about 3.5 arc seconds, which, at a distance of 4.4 light-years, means that this star moves 0.00007 light-year in one year. It thus has a projected velocity in the plane of the sky of 22 km per second. (One kilometre is about 0.62 mile.) As for motion along the line of sight, Alpha Centauri's spectral lines are slightly blueshifted, implying a velocity of approach of about 20 km per second. The true space motion, equal to $(22^2 + 20^2)^{1/2}$ or about 30 km per second, suggests that this star will make its closest approach to the Sun (at three light-years' distance) some 280 centuries from now.

## Light from the Stars

## Stellar Magnitudes

Stellar brightnesses are usually expressed by means of their magnitudes, a usage inherited from classical times. A star of the first magnitude is about 2.5 times as bright as one of the second magnitude, which in turn is some 2.5 times as bright as one of the third magnitude, and so on. A star of the first magnitude is therefore $2.5^5$ or 100 times as bright as one of the sixth magnitude. The magnitude of Sirius, which appears to an observer on Earth as the brightest star in the sky (save the Sun), is −1.4. Canopus, the second brightest, has a magnitude of −0.7, while the faintest star normally seen without the aid of a telescope is of the sixth magnitude. Stars as faint as the 30th magnitude have been measured with modern telescopes, meaning that these instruments can detect stars about four billion times fainter than can the human eye alone.

The scale of magnitudes comprises a geometric progression of brightness. Magnitudes can be converted to light ratios by letting $ln$ and $lm$ be the brightnesses of stars of magnitudes $n$ and $m$; the logarithm of the ratio of the two brightnesses then equals 0.4 times the difference between them—i.e., $\log(lm/ln) = 0.4(n - m)$. Magnitudes are actually defined in terms of observed brightness, a quantity that depends on the light-detecting device employed. Visual magnitudes were originally measured with the eye, which is most sensitive to yellow-green light, while photographic magnitudes were obtained from images on old photographic plates, which were most sensitive to blue light. Today, magnitudes are measured electronically, using detectors such as CCDs equipped with yellow-green or blue filters to create conditions that roughly correspond to those under which the original visual and photographic magnitudes were measured. Yellow-green magnitudes are still often designated $V$ magnitudes, but blue magnitudes are now designated $B$. The scheme has been extended to other magnitudes, such as ultraviolet ($U$), red ($R$), and near-infrared ($I$). Other systems vary the details of this scheme. All magnitude systems must have a reference, or zero, point. In practice, this is fixed arbitrarily by agreed-upon magnitudes measured for a variety of standard stars.

The actually measured brightnesses of stars give apparent magnitudes. These cannot be converted to intrinsic brightnesses until the distances of the objects concerned are known. The absolute magnitude of a star is defined as the magnitude it would have if it were viewed at a standard distance of 10 parsecs (32.6 light-years). Since the apparent visual magnitude of the Sun is −26.75, its absolute magnitude corresponds to a diminution in brightness by a factor of $(2{,}062{,}650)^2$ and is, using logarithms, −26.75 + 2.5 × $\log(2{,}062{,}650)^2$, or −26.75 + 31.57 = 4.82. This is the magnitude that the Sun would have if it were at a distance of 10 parsecs—an object still visible to the naked eye, though not a very conspicuous one and certainly not the brightest in the sky. Very luminous stars, such as Deneb, Rigel, and Betelgeuse, have absolute magnitudes of −7 to −9, while one of the faintest known stars, the companion to the star with the catalog name BD + 4°4048, has an absolute visual magnitude of +19, which is about a million times fainter than the Sun. Many astronomers suspect that large numbers of such faint stars exist, but most of these objects have so far eluded detection.

## Stellar Colours

Stars differ in colour. Most of the stars in the constellation Orion visible to the naked eye are blue-white, most notably Rigel (Beta Orionis), but Betelgeuse (Alpha Orionis) is a deep red. In the telescope, Albireo (Beta Cygni) is seen as two stars, one blue and the other orange. One quantitative means of measuring stellar colours involves a comparison of the yellow (visual) magnitude of the star with its magnitude measured through a blue filter. Hot, blue stars appear brighter through the blue filter, while the opposite is true for cooler, red stars. In all magnitude scales, one magnitude step corresponds to a brightness ratio of 2.512. The zero point is chosen so that white stars with surface temperatures of about 10,000 K have the same visual and blue magnitudes. The conventional colour index is defined as the blue magnitude, $B$, minus the visual magnitude, $V$; the colour index, $B - V$, of the Sun is thus $+5.47 - 4.82 = 0.65$.

Orion.

The constellation Orion is one of the easiest to recognize because of a group of three stars. The three stars form a straight line that is often called Orion's Belt. The Orion Nebula can be seen as a pink fuzzy light below the line of three stars. The red star Betelgeuse is in the upper left, and the bright star Rigel is in the lower right.

## Magnitude Systems

Problems arise when only one colour index is observed. If, for instance, a star is found to have, say, a B − V colour index of 1.0 (i.e., a reddish colour), it is impossible without further information to decide whether the star is red because it is cool or whether it is really a hot star whose colour has been reddened by the passage of light through interstellar dust. Astronomers have overcome these difficulties by measuring the magnitudes of the same stars through three or more filters, often U (ultraviolet), B, and V.

Observations of stellar infrared light also have assumed considerable importance. In addition, photometric observations of individual stars from spacecraft and rockets have made possible the measurement of stellar colours over a large range of wavelengths. These data are important for hot stars and for assessing the effects of interstellar attenuation.

## Bolometric Magnitudes

The measured total of all radiation at all wavelengths from a star is called a bolometric magnitude. The corrections required to reduce visual magnitudes to bolometric magnitudes are large for very cool stars and for very hot ones, but they are relatively small for stars such as the Sun. A determination of the true total luminosity of a star affords a measure of its actual energy output. When the energy radiated by a star is observed from Earth's surface, only that portion to which the energy detector is sensitive and that can be transmitted through the atmosphere is recorded. Most of the energy of stars like the Sun is emitted in spectral regions that can be observed from Earth's surface. On the other hand, a cool dwarf star with a surface temperature of 3,000 K has an energy maximum on a wavelength scale at 10000 angstroms (Å) in the far-infrared, and most of its energy cannot therefore be measured as visible light. (One angstrom equals $10^{-10}$ metre, or 0.1 nanometre.) Bright, cool stars can be observed at infrared wavelengths, however, with special instruments that measure the amount of heat radiated by the star. Corrections for the heavy absorption of the infrared waves by water and other molecules in Earth's air must be made unless the measurements are made from above the atmosphere.

The hotter stars pose more difficult problems, since Earth's atmosphere extinguishes all radiation at wavelengths shorter than 2900 Å. A star whose surface temperature is 20,000 K or higher radiates most of its energy in the inaccessible ultraviolet part of the electromagnetic spectrum. Measurements made with detectors flown in rockets or spacecraft extend the observable wavelength region down to 1000 Å or lower, though most radiation of distant stars is extinguished below 912 Å—a region in which absorption by neutral hydrogen atoms in intervening space becomes effective.

To compare the true luminosities of two stars, the appropriate bolometric corrections must first be added to each of their absolute magnitudes. The ratio of the luminosities can then be calculated.

## Stellar Spectra

A star's spectrum contains information about its temperature, chemical composition, and intrinsic luminosity. Spectrograms secured with a slit spectrograph consist of a sequence of images of the slit in the light of the star at successive wavelengths. Adequate spectral resolution (or dispersion) might show the star to be a member of a close binary system, in rapid rotation, or to have an extended atmosphere. Quantitative determination of its chemical composition then becomes possible. Inspection of a high-resolution spectrum of the star may reveal evidence of a strong magnetic field.

## Line Spectrum

Spectral lines are produced by transitions of electrons within atoms or ions. As the electrons move closer to or farther from the nucleus of an atom (or of an ion), energy in the form of light (or other radiation) is emitted or absorbed. The yellow D lines of sodium or the H and K lines of ionized calcium (seen as dark absorption lines) are produced by discrete quantum jumps from the lowest energy levels (ground states) of these atoms. The visible hydrogen lines (the so-called Balmer series), however, are produced by electron transitions within atoms in the second energy level (or first excited state), which lies well above the ground level in energy. Only at high temperatures are sufficient numbers of atoms maintained in this state by collisions, radiations, and so forth to permit an appreciable number of absorptions to occur. At the low surface temperatures of a red dwarf star, few electrons populate the second level of hydrogen, and thus the hydrogen lines are dim. By contrast, at very high temperatures—for instance, that of the surface of a blue giant star—the hydrogen atoms are nearly all ionized and therefore cannot absorb or emit any line radiation. Consequently, only faint dark hydrogen lines are observed. The characteristic features of ionized metals such as iron are often weak in such hotter stars because the appropriate electron transitions involve higher energy levels that tend to be more sparsely populated than the lower levels. Another factor is that the general "fogginess," or opacity, of the atmospheres of these hotter stars is greatly increased, resulting in fewer atoms in the visible stellar layers capable of producing the observed lines.

The continuous (as distinct from the line) spectrum of the Sun is produced primarily by the photodissociation of negatively charged hydrogen ions (H–)—i.e., atoms of hydrogen to which an extra electron is loosely attached. In the Sun's atmosphere, when H– is subsequently destroyed by photodissociation, it can absorb energy at any of a whole range of wavelengths and thus produce a continuous range of absorption of radiation. The main source of light absorption in the hotter stars is the photoionization of hydrogen atoms, both from ground level and from higher levels.

## Spectral Analysis

The physical processes behind the formation of stellar spectra are well enough understood to permit determinations of temperatures, densities, and chemical compositions of stellar atmospheres. The star studied most extensively is, of course, the Sun, but many others also have been investigated in detail.

The general characteristics of the spectra of stars depend more on temperature variations among the stars than on their chemical differences. Spectral features also depend on the density of the absorbing atmospheric matter, and density in turn is related to a star's surface gravity. Dwarf stars, with great surface gravities, tend to have high atmospheric densities; giants and supergiants, with low surface gravities, have relatively low densities. Hydrogen absorption lines provide a case in point. Normally, an undisturbed atom radiates a very narrow line. If its energy levels are perturbed by charged particles passing nearby, it radiates at a wavelength near its characteristic wavelength. In a hot gas, the range of disturbance of the hydrogen lines is very high, so that the spectral line radiated by the whole mass of gas is spread out considerably; the amount of blurring depends on the density of the gas in a known fashion. Dwarf stars such as Sirius show broad hydrogen features with extensive "wings" where the line fades slowly out into the background, while supergiant stars, with less-dense atmospheres, display relatively narrow hydrogen lines.

## Classification of Spectral Types

Most stars are grouped into a small number of spectral types. The Henry Draper Catalogue and the Bright Star Catalogue list spectral types from the hottest to the coolest stars. These types are designated, in order of decreasing temperature, by the letters O, B, A, F, G, K, and M. This group is supplemented by R- and N-type stars (today often referred to as carbon, or C-type, stars) and S-type stars. The R-, N-, and S-type stars differ from the others in chemical composition; also, they are invariably giant or supergiant stars. With the discovery of brown dwarfs—objects that form like stars but do not shine through thermonuclear fusion—the system of stellar classification has been expanded to include spectral types L, T, and Y.

The spectral sequence O through M represents stars of essentially the same chemical composition but of different temperatures and atmospheric pressures. This simple interpretation, put forward in the 1920s by the Indian astrophysicist Meghnad N. Saha, has provided the physical basis for all subsequent interpretations of stellar spectra. The spectral sequence is also a colour sequence: the O- and B-type stars are intrinsically the bluest and hottest; the M-, R-, N-, and S-type stars are the reddest and coolest.

In the case of cool stars of type M, the spectra indicate the presence of familiar metals, including iron, calcium, magnesium, and also titanium oxide molecules (TiO), particularly in the red and green parts of the spectrum. In the somewhat hotter K-type stars, the TiO features disappear, and the spectrum exhibits a wealth of metallic lines. A few especially stable fragments of molecules such as cyanogen (CN) and the hydroxyl radical (OH) persist in these stars and even in G-type stars such as the Sun. The spectra of G-type stars are dominated by the characteristic lines of metals, particularly those of iron, calcium, sodium, magnesium, and titanium.

The behaviour of calcium illustrates the phenomenon of thermal ionization. At low temperatures a calcium atom retains all of its electrons and radiates a spectrum characteristic of the neutral, or normal, atom; at higher temperatures collisions between atoms and electrons and the absorption of radiation both tend to detach electrons and to produce singly ionized calcium atoms. At the same time, these ions can recombine with electrons to produce neutral calcium atoms. At high temperatures or low electron pressures, or both, most of the atoms are ionized. At low temperatures and high densities, the equilibrium favours the neutral state. The concentrations of ions and neutral atoms can be computed from the temperature, the density, and the ionization potential (namely, the energy required to detach an electron from the atom).

The absorption line of neutral calcium at 4227 Å is thus strong in cool M-type dwarf stars, in which the pressure is high and the temperature is low. In the hotter G-type stars, however, the lines of ionized calcium at 3968 and 3933 Å (the H and K lines) become much stronger than any other feature in the spectrum.

In stars of spectral type F, the lines of neutral atoms are weak relative to those of ionized atoms. The hydrogen lines are stronger, attaining their maximum intensities in A-type stars, in which the surface temperature is about 9,000 K. Thereafter, these absorption lines gradually fade as the hydrogen becomes ionized.

The hot B-type stars, such as Epsilon Orionis, are characterized by lines of helium and of singly ionized oxygen, nitrogen, and neon. In very hot O-type stars, lines of ionized helium appear. Other prominent features include lines of doubly ionized nitrogen, oxygen, and carbon and of triply ionized silicon, all of which require more energy to produce.

In the more modern system of spectral classification, called the MK system (after the American astronomers William W. Morgan and Philip C. Keenan, who introduced it), luminosity class is assigned to the star along with the Draper spectral type. For example, the star Alpha Persei is classified as F5 Ib, which means that it falls about halfway between the beginning of type F (i.e., F0) and of type G (i.e., G0). The Ib suffix means that it is a moderately luminous supergiant. The star Pi Cephei, classified as G2 III, is a giant falling between G0 and K0 but much closer to G0. The Sun, a dwarf star of type G2, is classified as G2 V. A star of luminosity class II falls between giants and supergiants; one of class IV is called a subgiant.

## Bulk Stellar Properties

## Stellar Temperatures

Temperatures of stars can be defined in a number of ways. From the character of the spectrum and the various degrees of ionization and excitation found from its analysis, an ionization or excitation temperature can be determined.

A comparison of the V and B magnitudes yields a B − V colour index, which is related to the colour temperature of the star. The colour temperature is therefore a measure of the relative amounts of radiation in two more or less broad wavelength regions, while the ionization and excitation temperatures pertain to the temperatures of strata wherein spectral lines are formed.

Provided that the angular size of a star can be measured and that the total energy flux received at Earth (corrected for atmospheric extinction) is known, the so-called brightness temperature can be found.

The effective temperature, $T_{eff}$, of a star is defined in terms of its total energy output and radius. Thus, since $\sigma T^4_{eff}$ is the rate of radiation per unit area for a perfectly radiating sphere and if $L$ is the total radiation (i.e., luminosity) of a star considered to be a sphere of radius $R$, such a sphere (called a blackbody) would emit a total amount of energy equal to its surface area, $4\pi R^2$, multiplied by its energy per unit area. In symbols, $L = 4\pi R^2 \sigma T^4_{eff}$. This relation defines the star's equivalent blackbody, or effective, temperature.

Since the total energy radiated by a star cannot be directly observed (except in the case of the Sun), the effective temperature is a derived quantity rather than an observed one. Yet, theoretically, it is the fundamental temperature. If the bolometric corrections are known, the effective temperature can be found for any star whose absolute visual magnitude and radius are known. Effective temperatures are closely related to spectral type and range from about 40,000 K for hot O-type stars, through 5,800 K for stars like the Sun, to about 300 K for brown dwarfs.

## Stellar Masses

Masses of stars can be found directly only from binary systems and only if the scale of the orbits of the stars around each other is known. Binary stars are divided into three categories, depending on the mode of observation employed: visual binaries, spectroscopic binaries, and eclipsing binaries.

## Visual Binaries

Visual binaries can be seen as double stars with the telescope. True doubles, as distinguished from apparent doubles caused by line-of-sight effects, move through space together and display a common space motion. Sometimes a common orbital motion can be measured as well. Provided that the distance to the binary is known, such systems permit a determination of stellar masses, $m_1$ and $m_2$, of the two members. The angular radius, $a''$, of the orbit (more accurately, its semimajor axis) can be measured directly, and, with the distance known, the true dimensions of the semimajor axis, $a$, can be found. If $a$ is expressed in astronomical units, which is given by $a$ (measured in seconds of arc) multiplied by the distance in parsecs, and the period, $P$, also measured directly, is expressed in years, then the sum of the masses of the two orbiting stars can be found from an application of Kepler's third law. (An astronomical unit is the average distance from Earth to the Sun, 149,597,870.7 km [92,955,807.3 miles].) In symbols, $(m_1 + m_2) = a^3/P^2$ in units of the Sun's mass. For example, for the binary system 70 Ophiuchi, $P$ is 87.8 years, and the distance is 5.0 parsecs; thus, $a$ is 22.8 astronomical units, and $m_1 + m_2 = 1.56$ solar masses. From a measurement of the motions of the two members relative to the background stars, the orbit of each star has been determined with respect to their common centre of gravity. The mass ratio, $m_2/(m_1 + m_2)$, is 0.42; the individual masses for $m_1$ and $m_2$, respectively, are then 0.90 and 0.66 solar mass.

The star known as 61 Cygni was the first whose distance was measured (via parallax by the German astronomer Friedrich W. Bessel in the mid-19th century). Visually, 61 Cygni is a double star separated by 83.2 astronomical units. Its members move around one another with a period of 653 years. It was among the first stellar systems thought to contain a potential planet, although this has not been confirmed and is now considered unlikely. Nevertheless, since the 1990s a variety of discovery techniques have confirmed the existence of thousands of planets orbiting other stars.

## Spectroscopic Binaries

Spectroscopic binary stars are found from observations of radial velocity. At least the brighter member of such a binary can be seen to have a continuously changing periodic velocity that alters the wavelengths of its spectral lines in a rhythmic way; the velocity curve repeats itself exactly from one cycle to the next, and the motion can be interpreted as orbital motion. In some cases, rhythmic changes in the lines of both members can be measured. Unlike visual binaries, the semimajor axes or the individual masses cannot be found for most spectroscopic binaries, since the angle between the orbit plane and the plane of the sky cannot be determined. If spectra from both members are observed, mass ratios can be found. If one spectrum alone is observed, only a quantity called the mass function can be derived, from which is calculated a lower limit to the stellar masses. If a spectroscopic binary is also observed to be an eclipsing system, the inclination of the orbit and often the values of the individual masses can be ascertained.

## Eclipsing Binaries

An eclipsing binary consists of two close stars moving in an orbit so placed in space in relation to Earth that the light of one can at times be hidden behind the other. Depending on the orientation of the orbit and sizes of the stars, the eclipses can be total or annular (in the latter, a ring of one star shows behind the other at the maximum of the eclipse) or both eclipses can be partial. The best known example of an eclipsing binary is Algol (Beta Persei), which has a period (interval between eclipses) of 2.9 days. The brighter (B8-type) star contributes about 92 percent of the light of the system, and the eclipsed star provides less than 8 percent. The system contains a third star that is not eclipsed. Some 20 eclipsing binaries are visible to the naked eye.

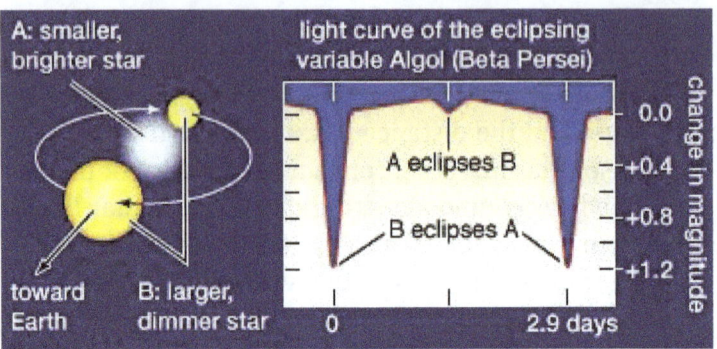

Light curve of Algol (Beta Persei), an eclipsing variable, or eclipsing binary, star system. The relative brightness of the system is plotted against time. A sharp dip occurs every 2.9 days when the fainter component star eclipses the brighter one, a shallower dip when the brighter star eclipses the fainter one.

The light curve for an eclipsing binary displays magnitude measurements for the system over a complete light cycle. The light of the variable star is usually compared with that of a nearby (comparison) star thought to be fixed in brightness. Often, a deep, or primary, minimum is produced when the component having the higher surface brightness is eclipsed. It represents the total eclipse and is characterized by a flat bottom. A shallower secondary eclipse occurs when the brighter component passes in front of the other; it corresponds to an annular eclipse (or transit). In a partial eclipse neither star is ever completely hidden, and the light changes continuously during an eclipse.

The shape of the light curve during an eclipse gives the ratio of the radii of the two stars and also one radius in terms of the size of the orbit, the ratio of luminosities, and the inclination of the orbital plane to the plane of the sky.

If radial-velocity curves are also available—i.e., if the binary is spectroscopic as well as eclipsing—additional information can be obtained. When both velocity curves are observable, the size of the orbit as well as the sizes, masses, and densities of the stars can be calculated. Furthermore, if the distance of the system is measurable, the brightness temperatures of the individual stars can be estimated from their luminosities and radii. All of these procedures have been carried out for the faint binary Castor C (two red-dwarf components of the six-member Castor multiple star system) and for the bright B-type star Mu Scorpii.

Close stars may reflect each other's light noticeably. If a small, high-temperature star is paired with a larger object of low surface brightness and if the distance between the stars is small, the part of the cool star facing the hotter one is substantially brightened by it. Just before (and just after) secondary eclipse, this illuminated hemisphere is pointed toward the observer, and the total light of the system is at a maximum.

The properties of stars derived from eclipsing binary systems are not necessarily applicable to isolated single stars. Systems in which a smaller, hotter star is accompanied by a larger, cooler object are easier to detect than are systems that contain, for example, two main-sequence stars. In such an unequal system, at least the cooler star has certainly been affected by evolutionary changes, and probably so has the brighter one. The evolutionary development of two stars near one another does not exactly parallel that of two well-separated or isolated ones.

Eclipsing binaries include combinations of a variety of stars ranging from white dwarfs to huge supergiants (e.g., VV Cephei), which would engulf Jupiter and all the inner planets of the solar system if placed at the position of the Sun.

Some members of eclipsing binaries are intrinsic variables, stars whose energy output fluctuates with time. In many such systems, large clouds of ionized gas swirl between the stellar members. In others, such as Castor C, at least one of the faint M-type dwarf components might be a flare star, one in which the brightness can unpredictably and suddenly increase to many times its normal value.

## Binaries and Extrasolar Planetary Systems

Near the Sun, most stars are members of binaries, and many of the nearest single stars are suspected of having companions. Although some binary members are separated by hundreds of astronomical units and others are contact binaries (stars close enough for material to pass between them), binary systems are most frequently built on the same scale as that of the solar system—namely, on the order of about 10 astronomical units. The division in mass between two components of a binary seems to be nearly random. A mass ratio as small as about 1:20 could occur about 5 percent of the time, and under these circumstances a planetary system comparable to the solar system is able to form.

The formation of double and multiple stars on the one hand and that of planetary systems on the other seem to be different facets of the same process. Planets are probably produced as a natural by-product of star formation. Only a small fraction of the original nebula matter is likely to be retained in planets, since much of the mass and angular momentum is swept out of the system. Conceivably, as many as 100 million stars could have bona fide planets in the Milky Way Galaxy.

Individual planets around other stars—i.e., extrasolar planets—are very difficult to observe directly because a star is always much brighter than its attendant planet. Jupiter, for example, would be only one-billionth as bright as the Sun and appear so close to it as to be undetectable from even the nearest star. If candidate stars are treated as possible spectroscopic binaries, however, then one may look for a periodic change in the star's radial velocity caused by a planet swinging around it. The effect is very small: even Jupiter would cause a change in the apparent radial velocity of the Sun of only about 10 metres (33 feet) per second spread over Jupiter's orbital period of about 12 years at best. Current techniques using very large telescopes to study fairly bright stars can measure radial velocities with a precision of about a metre per second, provided that the star has very sharp spectral lines, such as is observed for Sun-like stars and stars of types K and M. This means that at present the radial-velocity method normally can detect planets like Earth around M-type stars. Moreover, the closer the planet is to its parent star, the greater and quicker the velocity swing, so that detection of giant planets close to a star is favoured over planets farther out. Finally, even when a planet is detected, the usual spectroscopic binary problem of not knowing the angle between the orbit plane and that of the sky allows only a minimum mass to be assigned to the planet.

One exception to this last problem is HD 209458, a seventh-magnitude G0 V star about 150 light-years away with a planetary object orbiting it every 3.5 days. Soon after the companion, HD 209458b, was discovered in 1999 by its effect on the star's radial velocity, it also was found to be eclipsing the star, meaning that its orbit is oriented almost edge-on toward Earth. This type of eclipse is called a transit, and this method has been used, most notably by the Kepler satellite, to find thousands of extrasolar planets. Some of these planets are roughly the size of Earth and can be found in their star's habitable zone, the distance from a star where liquid water, and thus possibly life, can survive on the surface.

The first planet found to orbit two stars, Kepler-16b.

This transit of HD 209458b, as well as observations of spectral lines in the planet's atmosphere, allowed determination of the planet's mass and radius—0.64 and 1.38 times those of Jupiter, respectively. These numbers imply that the planet is even more of a giant than Jupiter itself. What was unexpected is its proximity to the parent star—more than 100 times closer than Jupiter is to the Sun, raising the question of how a giant gaseous planet that close can survive the star's radiation. The fact that many other extrasolar planets have been found to have orbital periods measured in days rather than years, and thus to be very close to their parent stars, suggests that the HD 209458 case is not unusual. There are also some confirmed cases of planets around supernova remnants called pulsars, although whether the planets preceded the supernova explosions that produced the pulsars or were acquired afterward remains to be determined.

The first extrasolar planets were discovered in 1992. Thousands of extrasolar planets are known, with more such discoveries being added regularly.

In addition to the growing evidence for existence of extrasolar planets, space-based observatories designed to detect infrared radiation have found many young stars (including Vega, Fomalhaut, and Beta Pictoris) to have disks of warm matter orbiting them. This matter is composed of myriad particles mostly about the size of sand grains and might be taking part in the first stage of planetary formation.

The infrared emission from the young star Fomalhaut and the dust belt surrounding it, as recorded with the European Space Agency's satellite observatory Herschel.

## Mass Extremes

The mass of most stars lies within the range of 0.3 to 3 solar masses. The star with the largest mass determined to date is R136a1, a giant of about 265 solar masses that had as much as 320 solar masses when it was formed. There is a theoretical upper limit to the masses of nuclear-burning stars (the Eddington limit), which limits stars to no more than a few hundred solar masses. On the low mass side, most stars seem to have at least 0.1 solar mass. The theoretical lower mass limit for an ordinary star is about 0.075 solar mass, for below this value an object cannot attain a central temperature high enough to enable it to shine by nuclear energy. Instead, it may produce a much lower level of energy by gravitational shrinkage. If its mass is not much below the critical 0.075 solar mass value, it will appear as a very cool, dim star known as a brown dwarf. Its evolution is simply to continue cooling toward eventual extinction. At still somewhat lower masses, the object would be a giant planet. Jupiter, with a mass roughly 0.001 that of the Sun, is just such an object, emitting a very low level of energy (apart from reflected sunlight) that is derived from gravitational shrinkage.

Brown dwarfs were late to be discovered, the first unambiguous identification having been made in 1995. It is estimated, however, that hundreds must exist in the solar neighbourhood. An extension of the spectral sequence for objects cooler than M-type stars has been constructed, using L for warmer brown dwarfs, T for cooler ones, and Y for the coolest. The presence of methane in the T brown dwarfs and of ammonia in the Y brown dwarfs emphasizes their similarity to giant planets.

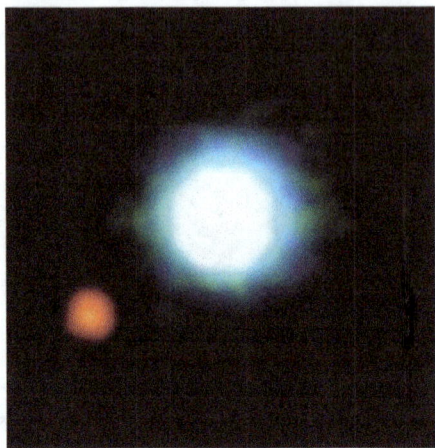

The brown dwarf 2MASSWJ 1207334–393254 (centre) as seen in a photo taken by the Very Large Telescope at the European Southern Observatory, Cerro Paranal, Chile. The brown dwarf has a mass 25 times that of Jupiter and a surface temperature of 2,400 K. Orbiting the brown dwarf at a distance of 8.3 billion km (5.2 billion miles) is a planet (lower left) that has a mass five times that of Jupiter and a surface temperature of 1,250 K.

## Stellar Radii

Angular sizes of bright red giant and supergiant stars were first measured directly during the 1920s, using the principle of interference of light. Only bright stars with large angular size can be measured by this method. Provided the distance to the star is known, the physical radius can be determined.

Eclipsing binaries also provide extensive data on stellar dimensions. The timing of eclipses provides the angular size of any occulting object, and so analyzing the light curves of eclipsing binaries can be a useful means of determining the dimensions of either dwarf or giant stars. Members of close binary systems, however, are sometimes subject to evolutionary effects, mass exchange, and other disturbances that change the details of their spectra.

A more recent method, called speckle interferometry, has been developed to reproduce the true disks of red supergiant stars and to resolve spectroscopic binaries such as Capella. The speckle phenomenon is a rapidly changing interference-diffraction effect seen in a highly magnified diffraction image of a star observed with a large telescope.

If the absolute magnitude of a star and its temperature are known, its size can be computed. The temperature determines the rate at which energy is emitted by each unit of area, and the total luminosity gives the total power output. Thus, the surface area of the star and, from it, the radius of the object can be estimated. This is the only way available for estimating the dimensions of white dwarf stars. The chief uncertainty lies in choosing the temperature that represents the rate of energy emission.

## Average Stellar Values

Main-sequence stars range from very luminous objects to faint M-type dwarf stars, and they vary considerably in their surface temperatures, their bolometric (total) luminosities, and their radii. Moreover, for stars of a given mass, a fair spread in radius, luminosity, surface temperature, and spectral type may exist. This spread is produced by stellar evolutionary effects and tends to broaden the main sequence. Masses are obtained from visual and eclipsing binary systems observed spectroscopically. Radii are found from eclipsing binary systems, from direct measurements in a few favourable cases, by calculations, and from absolute visual magnitudes and temperatures.

Average values for radius, bolometric luminosity, and mass are meaningful only for dwarf stars. Giant and subgiant stars all show large ranges in radius for a given mass. Conversely, giant stars of very nearly the same radius, surface temperature, and luminosity can have appreciably different masses.

## Stellar Statistics

Some of the most important generalizations concerning the nature and evolution of stars can be derived from correlations between observable properties and from certain statistical results. One of the most important of these correlations concerns temperature and luminosity—or, equivalently, colour and magnitude.

## Hertzsprung-Russell Diagram

When the absolute magnitudes of stars, or their intrinsic luminosities on a logarithmic scale, are plotted in a diagram against temperature or, equivalently, against the spectral types, the stars do not fall at random on the diagram but tend to congregate in certain restricted domains. Such a plot is usually called a Hertzsprung-Russell diagram, named for the early 20th-century astronomers Ejnar Hertzsprung of Denmark and Henry Norris Russell of the United States, who independently discovered the relations shown in it. As is seen in the diagram, most of the congregated stars are dwarfs lying closely around a diagonal line called the main sequence. These stars range from hot, O- and B-type, blue objects at least 10,000 times brighter than the Sun down through white A-type stars such as Sirius to orange K-type stars such as Epsilon Eridani and finally to M-type red dwarfs thousands of times fainter than the Sun. The sequence is continuous; the luminosities fall off smoothly with decreasing surface temperature; the masses and radii decrease but at a much slower rate; and the stellar densities gradually increase.

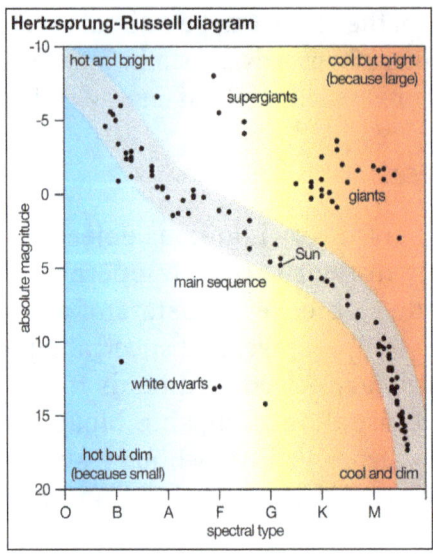

Hertzsprung-Russell diagram.

The second group of stars to be recognized was a group of giants—such objects as Capella, Arcturus, and Aldebaran—which are yellow, orange, or red stars about 100 times as bright as the Sun and have radii on the order of 10–30 million km (about 6–20 million miles, or 15–40 times as large as the Sun). The giants lie above the main sequence in the upper right portion of the diagram. The category of supergiants includes stars of all spectral types; these stars show a large spread in intrinsic brightness, and some even approach absolute magnitudes of −7 or −8. A few red supergiants, such as the variable star VV Cephei, exceed in size the orbit of Jupiter or even that of Saturn, although most of them are smaller. Supergiants are short-lived and rare objects, but they can be seen at great distances because of their tremendous luminosity.

Subgiants are stars that are redder and larger than main-sequence stars of the same luminosity. Many of the best-known examples are found in close binary systems where conditions favour their detection.

The white dwarf domain lies about 10 magnitudes below the main sequence. These stars are in the last stages of their evolution.

The spectrum-luminosity diagram has numerous gaps. Few stars exist above the white dwarfs and to the left of the main sequence. The giants are separated from the main sequence by a gap named for Hertzsprung, who in 1911 became the first to recognize the difference between main-sequence and giant stars. The actual concentration of stars differs considerably in different parts of the diagram. Highly luminous stars are rare, whereas those of low luminosity are very numerous.

The spectrum-luminosity diagram applies to the stars in the galactic spiral arm in the neighbourhood of the Sun and represents what would be obtained if a composite Hertzsprung-Russell diagram were constructed combining data for a large number of the star groups called open (or galactic) star clusters, as, for example, the double cluster h and χ Persei, the Pleiades, the Coma cluster, and the Hyades. It includes very young stars, a few million years old, as well as ancient stars perhaps as old as 10 billion years.

By contrast, another Hertzsprung-Russell diagram exhibits the type of temperature-luminosity, or colour-magnitude, relation characteristic of stars in globular clusters, in the central bulge of the Galaxy, and in elliptical external galaxies—namely, of the so-called stellar Population II. (In addition to these oldest objects, Population II includes other very old stars that occur between the spiral arms of the Galaxy and at some distance above and below the galactic plane.) Because these systems are very remote from the observer, the stars are faint, and their spectra can be observed only with difficulty. As a consequence, their colours rather than their spectra must be measured. Since the colours are closely related to surface temperature and therefore to spectral types, equivalent spectral types may be used, but it is stellar colours, not spectral types, that are observed in this instance.

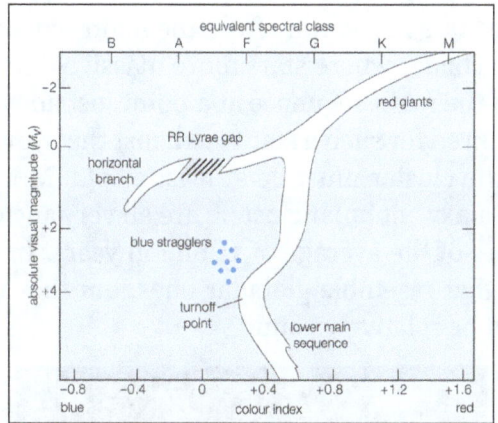

Colour-magnitude (Hertzsprung-Russell) diagram for an old globular cluster made up of Population II stars.

The differences between the two Hertzsprung-Russell diagrams are striking. In the second there are no supergiants, and, instead of a domain at an absolute magnitude of about 0, the giant stars form a branch that starts high and to the right at about −3.5 for very red stars and flows in a continuous sequence until it reaches an absolute magnitude of about 0. At that point the giant branch splits: a main band of stars, all about the same colour, proceeds downward (i.e., to fainter stars) to a magnitude of about +3 and then connects to the main sequence at about +4 by way of a narrow band. The main sequence of Population II stars extends downward to fainter, redder stars in much the same way as in the spiral-arm Population I stars. (Population I is the name given to the stars found within the spiral arms of the Milky Way system and other galaxies of the same type. Containing stars of all ages, from those in the process of formation to defunct white dwarfs, Population I is nonetheless always associated with the gas and dust of the interstellar medium.) The main sequence ends at about spectral type G, however, and does not extend up through the A, B, and O spectral types, though occasionally a few such stars, blue stragglers, are found in the region normally occupied by the main sequence. The blue stragglers are likely red stars that have gained mass from another star through stellar collision or mass transfer.

The other band of stars formed from the split of the giant branch is the "horizontal branch," which falls near magnitude +0.6 and fills the aforementioned Hertzsprung gap, extending to increasingly blue stars beyond the RR Lyrae stars, which are indicated by the crosshatched area in the diagram. Among these blue hot stars are found novas and the nuclei of planetary nebulas, the latter so called because their photographic image resembles that of a distant planet. Not all globular clusters show identical colour-magnitude diagrams, which may be due to differences in the cluster ages or other factors.

## Estimates of Stellar Ages

The shapes of the colour-magnitude diagrams permit estimates of globular-cluster ages. The point at which stars move away from the main sequence is called the turnoff point, For example, in a cluster where stars more massive than about 1.3 solar masses have evolved away from the main sequence at a point just above the position occupied by the Sun, the time required for such a star to exhaust the hydrogen in its core is about 5–6 billion years, and the cluster must be at least as old. More ancient clusters have been identified. In the Galaxy, globular clusters are all very ancient objects, having ages within a few billion years of the average of 11 billion years. In the Magellanic Clouds, however, clusters exist that resemble globular ones, but they contain numerous blue stars and therefore must be relatively young.

Infant stars in the Small Magellanic Cloud.

Open clusters in the spiral arms of the Galaxy—extreme Population I—tell a somewhat different story. A colour-magnitude diagram can be plotted for a number of different open clusters—for example, the double cluster h and χ Persei, the Pleiades, Praesepe, and M67—with the main feature distinguishing the clusters being their ages. The young cluster h and χ Persei, which is a few million years old, contains stars ranging widely in luminosity. Some stars have already evolved into the supergiant stage (in such a diagram the top of the main sequence is bent over). The stars of luminosity 10,000 times greater than that of the Sun have already largely depleted the hydrogen in their cores and are leaving the main sequence.

The brightest stars of the Pleiades cluster, aged about 100 million years, have begun to leave the main sequence and are approaching the critical phase when they will have exhausted all the hydrogen in their cores. There are no giants in the Pleiades. Presumably, the cluster contained no stars as massive as some of those found in h and χ Persei.

Pleiades: Bright nebulosity in the Pleiades (M45, NGC 1432), distance 490 light-years. Cluster stars provide the light, and surrounding clouds of dust reflect and scatter the rays from the stars.

The cluster known as Praesepe, or the Beehive, at an age of 790 million years, is older than the Pleiades. All stars much more luminous than the first magnitude have begun to leave the main sequence; there are some giants. The Hyades, about 620 million years old, displays a similar colour-magnitude array. These clusters contain a number of white dwarfs, indicating that the initially most luminous stars have already run the gamut of evolution. In a very old cluster such as M67, which is 4.5 billion years old, all of the bright main-sequence stars have disappeared.

The colour-magnitude diagrams for globular and open clusters differ quantitatively because the latter show a wider range of ages and differ in chemical composition. Most globular clusters have smaller metal-to-hydrogen ratios than do open clusters or the Sun. The gaps between the red giants and the blue main-sequence stars of the open clusters (Population I) often contain unstable stars such as variables. The Cepheid variable stars, for instance, fall in these gaps.

Cepheid variables, as seen by the Hubble Space Telescope.

The giant stars of the Praesepe cluster are comparable to the brightest stars in M67. The M67 giants have evolved from the main sequence near an absolute magnitude of +3.5, whereas the Praesepe giants must have masses about twice as great as those of the M67 giants. Giant stars of the same luminosity may therefore have appreciably different masses.

## Numbers of Stars Versus Luminosity

Of great statistical interest is the relationship between the luminosities of the stars and their frequency of occurrence. The naked-eye stars are nearly all intrinsically brighter than the Sun, but the opposite is true for the known stars within 20 light-years of the Sun. The bright stars are easily seen at great distances; the faint ones can be detected only if they are close.

The luminosity function (the number of stars with a specific luminosity) depends on population type. The luminosity function for pure Population II differs substantially from that for pure Population I. There is a small peak near absolute magnitude +0.6, corresponding to the horizontal branch for Population II, and no stars as bright as absolute magnitude –5. The luminosity function for pure Population I is evaluated best from open star clusters, the stars in such a cluster being at about the same distance. The neighbourhood of the Sun includes examples of both Populations I and II.

## Mass-luminosity Correlations

A plot of mass against bolometric luminosity for visual binaries for which good parallaxes and masses are available shows that for stars with masses comparable to that of the Sun the luminosity, $L$, varies as a power, $3 + \beta$, of the mass $M$. This relation can be expressed as $L = (M)^{3+\beta}$. The power differs for substantially fainter or much brighter stars.

This mass-luminosity correlation applies only to unevolved main-sequence stars. It fails for giants and supergiants and for the subgiant (dimmer) components of eclipsing binaries, all of which have changed considerably during their lifetimes. It does not apply to any stars in a globular cluster not on the main sequence, or to white dwarfs that are abnormally faint for their masses.

The mass-luminosity correlation, predicted theoretically in the early 20th century by the English astronomer Arthur Eddington, is a general relationship that holds for all stars having essentially the same internal density and temperature distributions—i.e., for what are termed the same stellar models.

## Variable Stars

Many stars are variable. Some are geometric variables, as in the eclipsing binaries considered earlier. Others are intrinsically variable—i.e., their total energy output fluctuates with time.

A fair number of stars are intrinsically variable. Some objects of this type were found by accident, but many were detected as a result of carefully planned searches. Variable stars are important in astronomy for several reasons. They usually appear to be stars at critical or short-lived phases of their evolution, and detailed studies of their light and spectral characteristics, spatial distribution, and association with other types of stars may provide valuable clues to the life histories of various classes of stars. Certain kinds of variable stars, such as Cepheids (periodic variables) and novas and supernovas (explosive variables), are extremely important in that they make it possible to establish the distances of remote stellar systems beyond the Galaxy. If the intrinsic luminosity of a recognizable variable is known and this kind of variable star can be found in a distant stellar system, the distance of the latter can be estimated from a measurement of apparent and absolute magnitudes, provided that the interstellar absorption is also known.

## Classification

Variables are often classified as behaving like a prototype star, and the entire class is then named for this star—e.g., RR Lyrae stars are those whose variability follows the pattern of the star RR Lyrae. The most important classes of intrinsically variable stars are the following:

- Pulsating variables—stars whose variations in light and colour are thought to arise primarily from stellar pulsations. These include Beta Canis Majoris stars, RR Lyrae stars, and Delta Scuti stars, all with short regular periods of less than a day; Cepheids, with periods between 1 and 100 days; and long-period variables, semiregular variables, and irregular red variables, usually with unstable periods of hundreds of days.

- Explosive, or catastrophic, variables—stars in which the variations are produced by the wrenching away of part of the star, usually the outer layers, in some explosive process. They include SS Cygni or U Geminorum stars, novas, and supernovas (the last of which are enormous explosions involving most of the matter in a star).

- Miscellaneous and special types of variables—R Coronae Borealis stars, T Tauri stars, flare stars, pulsars (neutron stars), spectrum and magnetic variables, X-ray variable stars, and radio variable stars.

## Pulsating Stars

An impressive body of evidence indicates that stellar pulsations can account for the variability of Cepheids, long-period variables, semiregular variables, Beta Canis Majoris stars, and even the irregular red variables. Of this group, the Cepheid variables have been studied in greatest detail, both theoretically and observationally. These stars are regular in their behaviour; some repeat their light curves with great faithfulness from one cycle to the next over periods of many years.

Much confusion existed in the study of Cepheids until it was recognized that different types of Cepheids are associated with different groups, or population types, of stars. Cepheids belonging to the spiral-arm Population I (or Type I Cepheids) are characterized by regularity in their behaviour. They show continuous velocity curves indicative of regular pulsation. They exhibit a relation between period and luminosity in the sense that the longer the period of the star, the greater its intrinsic brightness. This period-luminosity relationship has been used to establish the distances of remote stellar systems.

Cepheids with different properties are found in Population II, away from the Milky Way, in globular clusters. These Type II Cepheids are bluer than classic Population I Cepheids having the same period, and their light curves have different shapes. Studies of the light and velocity curves indicate that shells of gas are ejected from the stars as discontinuous layers that later fall back toward the surface. These stars exhibit a relation between period and luminosity different from that for Population I Cepheids, and thus the distance of a Cepheid in a remote stellar system can be determined only if its population type is known.

Closely associated with Population II Cepheids are the cluster-type, or RR Lyrae, variables. Many of these stars are found in clusters, but some, such as the prototype RR Lyrae, occur far from any cluster or the central galactic bulge. Their periods are less than a day, and there is no correlation between period and luminosity. Their absolute magnitudes are about 0.6, but somewhat dependent on metal abundance. They are thus about 50 times as bright as the Sun and so are useful for determining the distance of star clusters and some of the nearer external galaxies, their short periods permitting them to be detected readily.

Long-period variable stars also probably owe their variations to pulsations. Here the situation is complicated by the vast extent of their atmospheres, so that radiation originating at very different depths in the star is observed at the same time. At certain phases of the variations, bright hydrogen lines are observed, overlaid with titanium oxide absorption. The explanation is an outward-moving layer of hot, recombining gas, whose radiation is absorbed by strata of cool gases. These stars are all cool red giants and supergiants of spectral types M (normal composition), R and N (carbon-rich), or S (heavy-metal-rich). The range in visual brightness during a pulsation can be 100-fold, but the range in total energy output is much less, because at very low stellar temperatures (1,500–3,000 K) most of the energy is radiated in the infrared as heat rather than as light.

Unlike the light curves of classic Cepheids, the light curves of these red variables show considerable variations from one cycle to another. The visual magnitude of the variable star Mira Ceti (Omicron Ceti) is normally about 9–9.5 at minimum light, but at maximum it may lie between 5 and 2. Time intervals between maxima often vary considerably. In such cool objects, a very small change in temperature can produce a huge change in the output of visible radiation. At the low temperatures of the red variables, compounds and probably solid particles are formed copiously, so that the visible light may be profoundly affected by a slight change in physical conditions. Random fluctuations from cycle to cycle, which would produce negligible effects in a hotter star, produce marked light changes in a long-period variable.

Long-period variables appear to fall into two groups; those with periods of roughly 200 days tend to be associated with Population II, and those of periods of about a year belong to Population I.

Red semiregular variables such as the RV Tauri stars show complex light and spectral changes. They do not repeat themselves from one cycle to the next; their behaviour suggests a simultaneous operation of two or more modes of oscillation. Betelgeuse is an example of an irregular red variable. In these stars the free period of oscillation does not coincide with the periodicity of the driving mechanism.

Finally, among the various types of pulsating variable stars, the Beta Canis Majoris variables are high-temperature stars (spectral type B) that often show complicated variations in spectral-line shapes and intensities, velocity curves, and light. In many cases, they have two periods of variation so similar in duration that complex interference or beat phenomena are observed, both in radial velocities and in the shapes of spectral lines.

A large body of evidence suggests that all members of this first class of variable stars owe their variability to pulsation. The pulsation theory was first proposed as a possible explanation as early as 1879, was applied to Cepheids in 1914, and was further developed by Arthur Eddington in 1917–18. Eddington found that if stars have roughly the same kind of internal structure, then the period multiplied by the square root of the density equals a constant that depends on the internal structure.

The Eddington theory, though a good approximation, encountered some severe difficulties that have been met through modifications. If the entire star pulsated in synchronism, it should be brightest when compressed and smaller while faintest when expanded and at its largest. The radial velocity should be zero at both maximum and minimum light. Observations contradict these predictions. When the star pulsates, all parts of the main body move in synchronism, but the outer observable strata fall out of step or lag behind the pulsation of the inner regions. Pulsations involve only the outer part of a star; the core, where energy is generated by thermonuclear reactions, is unaffected.

Careful measurements of the average magnitudes and colours of RR Lyrae stars in the globular cluster M3 showed that all these stars fell within a narrow range of luminosity and colour (or surface temperature) or, equivalently, luminosity and radius. Also, every star falling in this narrow range of brightness and size was an RR Lyrae variable. Subsequent work has indicated that similar considerations apply to most classic Cepheids. Variability is thus a characteristic of any star whose evolution carries it to a certain size and luminosity, although the amplitude of the variability can vary dramatically.

In the pulsation theory as now developed, the light and velocity changes of Cepheids can be interpreted not only qualitatively but also quantitatively. The light curves of Cepheids, for example, have been precisely predicted by the theory. Stellar pulsation, like other rhythmic actions, may give rise to harmonic phenomena wherein beats reinforce or interfere with one another. Beat and interference phenomena then complicate the light and velocity changes. The RR Lyrae stars supply some of the best examples, but semiregular variables such as the RV Tauri stars or most Delta Scuti stars evidently vibrate simultaneously with two or more periods.

## Explosive Variables

The evolution of a member of a close double-star system can be markedly affected by the presence of its companion. As the stars age, the more massive one swells up more quickly as it moves away from the main sequence. It becomes so large that its outer envelope falls under the gravitational influence of the smaller star. Matter is continuously fed from the more rapidly evolving star to the less massive one, which still remains on the main sequence. U Cephei is a classic example of such a system for which spectroscopic evidence shows streams of gas flowing from the more highly evolved star to the hotter companion, which is now the more massive of the two. Eventually, the latter will also leave the main sequence and become a giant star, only to lose its outer envelope to the companion, which by that time may have reached the white dwarf stage.

Novas appear to be binary stars that have evolved from contact binaries of the W Ursae Majoris type, which are pairs of stars apparently similar to the Sun in size but revolving around one another while almost touching. One member may have reached the white dwarf stage. Matter fed to it from its distended companion appears to produce instabilities that result in violent explosions or nova outbursts. The time interval between outbursts can range from a few score years to hundreds of thousands of years.

In ordinary novas the explosion seems to involve only the outer layers, as the star later returns to its former brightness; in supernovas the explosion is catastrophic. Normally, novas are small blue stars much fainter than the Sun, though very much hotter. When an outburst occurs, the star can brighten very rapidly, by 10 magnitudes or more in a few hours. Thereafter it fades; the rate of fading is connected with the brightness of the nova. The brightest novas, which reach absolute magnitudes of about −10, fade most rapidly, whereas a typical slow nova, which reaches an absolute magnitude of −5, can take 10 or 20 times as long to decline in brightness. This property, when calibrated as the absolute magnitude at maximum brightness versus the time taken to decline by two magnitudes, allows novas to be used as distance indicators for nearby galaxies. The changes in light are accompanied by pronounced spectroscopic changes that can be interpreted as arising from alterations in an ejected shell that dissipates slowly in space. In its earliest phases, the expanding shell is opaque. As its area grows, with a surface temperature near 7,000 K, the nova brightens rapidly. Then, near maximum light, the shell becomes transparent, and its total brightness plummets rapidly, causing the nova to dim.

The mass of the shell is thought to be rather small, about 10–100 times the mass of Earth. Only the outer layers of the star seem to be affected; the main mass settles down after the outburst into a state much as before until a new outburst occurs. The existence of repeating novas, such as the star T Coronae Borealis, suggests that perhaps all novas repeat at intervals ranging up to thousands or perhaps millions of years; and probably, the larger the explosion, the longer the interval. There is strong evidence that novas are components of close double stars and, in particular, that they have evolved from the most common kind of eclipsing binaries, those of the W Ursae Majoris type.

Stars of the SS Cygni type, also known as dwarf novas, undergo novalike outbursts but of a much smaller amplitude. The intervals between outbursts are a few months to a year. Such variables are close binaries. The development of this particular type may be possible only in close binary systems.

There are two major types of supernovas, designated Type I (or SNe I) and Type II (or SNe II). They can be distinguished by the fact that Type II supernovas have hydrogen features in their spectra, while Type I supernovas do not. Type II supernovas arise from the collapse of a single star more massive than about eight solar masses, resulting in either a neutron star or a black hole. There are three classes of Type I supernovas: Ia, Ib, and Ic. Type Ia supernovas are believed to originate in a binary system containing a white dwarf, rather like the case of ordinary novas. Unlike the latter, however, in which only the outer layers of the white dwarf seem to be affected, in a Type I supernova the white dwarf is probably completely destroyed; the details are not yet fully understood. Certainly, a supernova's energy output is enormously greater than that of an ordinary nova. Type Ib and Ic supernovas are like Type II in that they are each the core collapse of a massive star. However, a Type II supernova retains its hydrogen envelope, whereas the Type Ib and Ic supernovas do not, thus leading to the different hydrogen features in their spectra. Type Ib retains a helium shell and so has a spectrum rich in helium lines; Type Ic does not retain the hydrogen or helium shell.

Supernova 1987A in the Large Magellanic Cloud.

Empirical evidence indicates that in a Type Ia supernova the absolute magnitude at maximum light can be determined by a combination of data derived from the rate of dimming after maximum, the shape of the light curve, and certain colour measurements. A comparison of the absolute and apparent magnitudes of maximum light in turn allows the distance of the supernova to be found. This is a matter of great usefulness because Type Ia supernovas at maximum light are the most luminous "standard candles" available for determining distances to external galaxies and thus can be observed in more distant galaxies than any other kind of standard candle. In 1999, application of this technique led to the totally unexpected discovery that the expansion of the universe is accelerating rather than slowing down. This acceleration is caused by dark energy, a gravitationally repulsive force that is the dominant component (73 percent) of the universe's mass-energy.

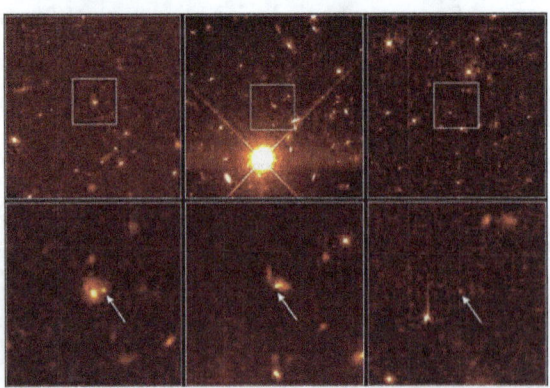

Supernova.

Three distant Type Ia supernovas, as observed by the Hubble Space Telescope in 1997. The bottom images are details of the upper wide views. The supernovas at left and centre occurred about five billion years ago, the right seven billion years ago.

A peculiar explosive variable with no known counterpart is Eta Carinae (NGC 3372), which appears in telescopes on Earth as a fuzzy red "star" slightly less than two seconds of arc in diameter. Surrounding it is a shell of gas and dust shaped roughly like an hourglass divided by a thin disk. First observed as a star of about the fourth magnitude in 1677, it brightened irregularly, undergoing an outburst in 1843, when it became for a few years the second brightest star in the sky. Thereafter it slowly faded, becoming too faint for the unaided eye around the turn of the 20th century. The fading was due, at least in part, to obscuration by dust emitted in the earlier eruption. The star remained near seventh magnitude with irregular variations for most of the 20th century, but it began brightening again by one or two tenths of a magnitude per year in the mid-1990s. In 2005 astronomers found that Eta Carinae is, in fact, a binary star system with an orbital period of 5.52 years. Its A component has a temperature of about 15,000 K; its B component, about 35,000 K. Eta Carinae is considered to be one of a small class of stars known as luminous blue variables. Its luminosity has been estimated as five million times that of the Sun. Flaring events producing not only visible effects but also X-ray, ultraviolet, and radio-wave effects have been observed.

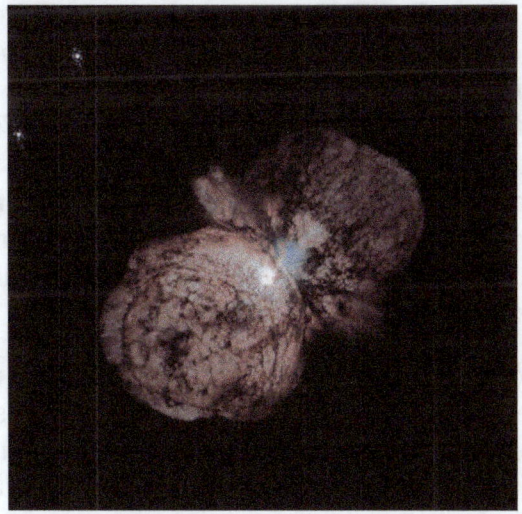

Eta Carinae: The hourglass and disk shapes of the gas and dust clouds,
are visible in this enhanced, computer-manipulated composite of eight
images taken by the Hubble Space Telescope.

Probably all variable stars represent more or less ephemeral phases in the evolution of a star. Aside from catastrophic events of the kind that produce a supernova, some phases of stellar variability might be of such brief duration as to permit recognizable changes during an interval of 50–100 years. Other stages may require many thousands of years. For example, the period of Delta Cephei, the prototype star of the Cepheid variables, has barely changed by a detectable amount since its variability was discovered in 1784.

## Peculiar Variables

R Coronae Borealis variables are giant stars of about the Sun's temperature whose atmospheres are characterized by excessive quantities of carbon and very little hydrogen. The brightness of such a star remains constant until the star suddenly dims by several magnitudes and then slowly recovers its original brightness. (The star's colour remains the same during the changes in brightness.) The dimmings occur in a random fashion and seem to be due to the huge concentrations of carbon. At times the carbon vapour literally condenses into soot, and the star is hidden until the smog blanket is evaporated. Similar veiling may sometimes occur in other types of low-temperature stars, particularly in long-period variables.

Flare stars are cool dwarfs (spectral type M) that display flares apparently very much like, but much more intense than, those of the Sun. In fact, the flares are sometimes so bright that they overwhelm the normal light of the star. Solar flares are associated with copious emission of radio waves, and simultaneous optical and radio-wave events appear to have been found in the stars UV Ceti, YZ Canis Minoris, and V371 Orionis.

Spectrum and magnetic variables, mostly of spectral type A, show only small amplitudes of light variation but often pronounced spectroscopic changes. Their spectra typically show strong lines of metals such as manganese, titanium, iron, chromium, and the lanthanides (also called rare earths), which vary periodically in intensity. These stars have strong magnetic fields, typically from a few hundred to a few thousand gauss. One star, HD 215441, has a field on the order of 30,000 gauss. (Earth's magnetic field has an average strength of about 0.5 gauss.) Not all magnetic stars are known to be variable in light, but such objects do seem to have variable magnetic fields. The best interpretation is that these stars are rotating about an inclined axis. As with Earth, the magnetic and rotation axes do not coincide. Different ions are concentrated in different areas (e.g., chromium in one area and the lanthanides in another).

The Sun is an emitter of radio waves, but, with present techniques, its radio emission could only just be detected from several parsecs away. Most discrete radio-frequency sources have turned out to be objects such as old supernovas, radio galaxies, or quasars, though well-recognized radio stars also have been recorded on occasion. These include flare stars, red supergiants such as Betelgeuse, the high-temperature dwarf companion to the red supergiant Antares, and the shells ejected from Nova Serpentis 1970 and Nova Delphini. The radio emission from the latter objects is consistent with that expected from an expanding shell of ionized gas that fades away as the gas becomes attenuated. The central star of the Crab Nebula has been detected as a radio (and optical) pulsar.

Measurements from rockets, balloons, and spacecraft have revealed distinct X-ray sources outside the solar system. The strongest galactic source, Scorpius X-1, appears to be associated with a hot variable star resembling an old nova. In all likelihood this is a binary star system containing a low-mass normal star and a nonluminous companion.

A number of globular clusters are sources of cosmic X-rays. Some of this X-ray emission appears as intense fluctuations of radiation lasting only a few seconds but changing in strength by as much as 25 times. These X-ray sources have become known as bursters, and several such objects have been discovered outside of globular clusters as well. Some bursters vary on a regular basis, while others seem to turn on and off randomly. The most popular interpretation holds that bursters are the result of binary systems in which one of the objects—a compact neutron star or black hole—pulls matter from the companion, a normal star. This matter is violently heated in the process, giving rise to X-rays. That the emission is often in the form of a burst is probably caused by something interrupting the flow of matter onto (or into) the compact object or by an eclipsing orbit of the binary system.

## Galaxy

Galaxies are titanic swarms of tens of millions to trillions of stars, orbiting around their common center of gravity. They also contain interstellar gas and dust. Galaxies show a range of shapes that astronomers group into three basic classes: spiral, elliptical, and irregular.

Spiral galaxies have three visible parts: a thin disk composed of stars, gas, and dust; a central bulge of older stars; and a spherical halo of the oldest stars and massive star clusters.

Spiral galaxy.

Elliptical galaxies have smooth, rounded shapes because the orbits of their stars are oriented in all directions. They contain little gas and dust, and no young stars. Like spiral galaxies, elliptical galaxies are surrounded by globular star clusters and dark matter.

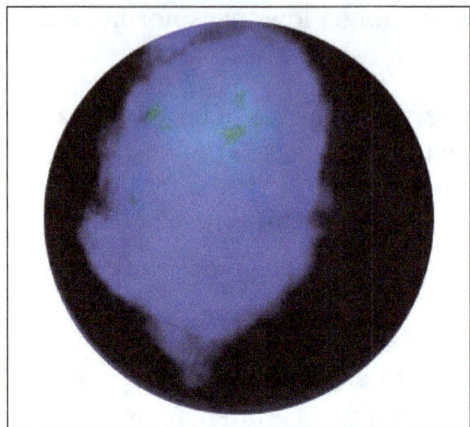

Elliptical galaxy.

Irregular galaxies have a chaotic appearance, and are usually small. Their irregular shapes are probably due to recent disturbances — either bursts of internal star formation, or gravitational encounters with external galaxies.

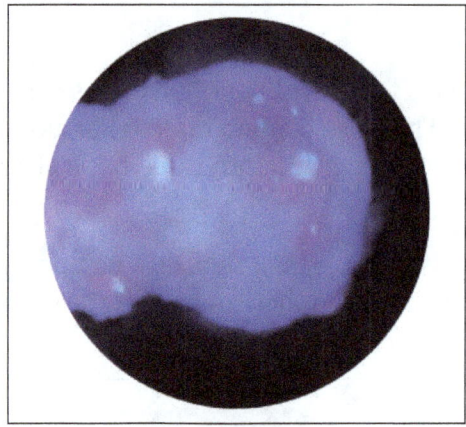

Irregular galaxy.

## The Milky Way

Our Milky Way is a barred spiral galaxy containing over 100 billion stars. The spiral disk of stars, gas, and dust is about 100,000 light-years across and 2,000 light-years thick — flatter than a pancake. The central bulge of stars is elongated in the shape of a bar.

The Sun orbits within the disk on the trailing edge of a minor spiral arm, about halfway between the galactic center and the visible edge. The Sun takes approximately 220 million years to circle the galaxy, and it has completed about 20 orbits since the solar system was born.

The Milky Way Galaxy resides in a neighborhood of a few dozen galaxies called the Local Group. They range in size from small dwarf galaxies to the large Andromeda Galaxy. Over time, these galaxies interact with one another, changing their motions and shapes. The long-term evolution of a galaxy is influenced by being part of a group. The Milky Way and the Andromeda Galaxy, our nearest spiral neighbor, are headed toward each other. In about five billion years, they may collide and merge. Eventually, our remote descendants could be living in a large elliptical galaxy.

Milky Way Galaxy.

## Nebula

The "Pillars of Creation" from the Eagle Nebula. Evidence from the Spitzer Telescope suggests that the pillars may already have been destroyed by a supernova explosion, but the light showing us the destruction will not reach the Earth for another millennium.

A nebula is an interstellar cloud of dust, hydrogen, helium and other ionized gases. Originally, the term was used to describe any diffuse astronomical object, including galaxies beyond the Milky Way. The Andromeda Galaxy, for instance, was once referred to as the *Andromeda Nebula* (and spiral galaxies in general as "spiral nebulae") before the true nature of galaxies was confirmed in the early 20th century by Vesto Slipher, Edwin Hubble and others.

Most nebulae are of vast size; some are hundreds of light-years in diameter. A nebula that is barely visible to the human eye from Earth would appear larger, but no brighter, from close by. The Orion Nebula, the brightest nebula in the sky and occupying an area twice the diameter of the full Moon, can be viewed with the naked eye but was missed by early astronomers. Although denser than the space surrounding them, most nebulae are far less dense than any vacuum created on Earth – a nebular cloud the size of the Earth would have a total mass of only a few kilograms. Many nebulae are visible due to fluorescence caused by embedded hot stars, while others are so diffuse they can only be detected with long exposures and special filters. Some nebulae are variably illuminated by T Tauri variable stars. Nebulae are often star-forming regions, such as in the "Pillars of Creation" in the Eagle Nebula. In these regions the formations of gas, dust, and other materials "clump" together to form denser regions, which attract further matter, and eventually will become dense enough to form stars. The remaining material is then believed to form planets and other planetary system objects.

## Formation

The Triangulum Emission Garren Nebula NGC 604.

There are a variety of formation mechanisms for the different types of nebulae. Some nebulae form from gas that is already in the interstellar medium while others are produced by stars. Examples of the former case are giant molecular clouds, the coldest, densest phase of interstellar gas, which can form by the cooling and condensation of more diffuse gas. Examples of the latter case are planetary nebulae formed from material shed by a star in late stages of its stellar evolution.

Star-forming regions are a class of emission nebula associated with giant molecular clouds. These form as a molecular cloud collapses under its own weight, proceeding stars. Massive stars may form in the center, and their ultraviolet radiation ionizes the surrounding gas, making it visible at optical wavelengths. The region of ionized hydrogen surrounding the massive stars is known as an H II region while the shells of neutral hydrogen surrounding the H II region are known as photodissociation region. Examples of star-forming regions are the Orion Nebula, the Rosette Nebula and the Omega Nebula. Feedback from star-formation, in the form of supernova explosions of massive stars, stellar winds or ultraviolet radiation from massive stars, or outflows from low-mass stars may disrupt the cloud, destroying the nebula after several million years.

Other nebulae form as the result of supernova explosions; the death throes of massive, short-lived stars. The materials thrown off from the supernova explosion are then ionized by the energy and the compact object that its core produces. One of the best examples of this is the Crab Nebula, in Taurus. The supernova event was recorded in the year 1054 and is labeled SN 1054. The compact object that was created after the explosion lies in the center of the Crab Nebula and its core is now a neutron star.

Still other nebulae form as planetary nebulae. This is the final stage of a low-mass star's life, like Earth's Sun. Stars with a mass up to 8–10 solar masses evolve into red giants and slowly lose their outer layers during pulsations in their atmospheres. When a star has lost enough material, its temperature increases and the ultraviolet radiation it emits can ionize the surrounding nebula that it has thrown off. Our Sun will produce a planetary nebula and its core will remain behind in the form of a white dwarf.

## Types of Nebulae

The Omega Nebula, an example of an emission nebula.

The Horsehead Nebula, an example of a dark nebula.

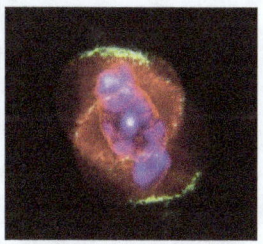

The Cat's Eye Nebula, an example of a planetary nebula.

The Red Rectangle Nebula, an example of a protoplanetary nebula.

The delicate shell of SNR B0509-67.5.

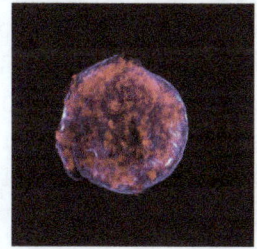

Tycho Supernova remnant in X-ray light.

## Classical Types

Objects named nebulae belong to 4 major groups. Before their nature was understood, galaxies ("spiral nebulae") and star clusters too distant to be resolved as stars were also classified as nebulae, but no longer are.

- H II regions, large diffuse nebulae containing ionized hydrogen

- Planetary nebulae

- Supernova remnant (e.g., Crab Nebula)

- Dark nebula

Not all cloud-like structures are named nebulae; Herbig–Haro objects are an example.

## Diffuse Nebulae

The Carina Nebula is a diffuse nebula.

Most nebulae can be described as diffuse nebulae, which means that they are extended and contain no well-defined boundaries. Diffuse nebulae can be divided into emission nebulae, reflection nebulae and dark nebulae. Visible light nebulae may be divided into emission nebulae that emit spectral line radiation from excited or ionized gas (mostly ionized hydrogen); they are often called H II regions (the term "H II" refers to ionized hydrogen). Reflection nebulae are visible primarily due to the light they reflect. Reflection nebulae themselves do not emit significant amounts of visible light, but are near stars and reflect light from them. Similar nebulae not illuminated by stars do not exhibit visible radiation, but may be detected as opaque clouds blocking light from luminous objects behind them; they are called dark nebulae.

Although these nebulae have different visibility at optical wavelengths, they are all bright sources of infrared emission, chiefly from dust within the nebulae.

## Planetary Nebulae

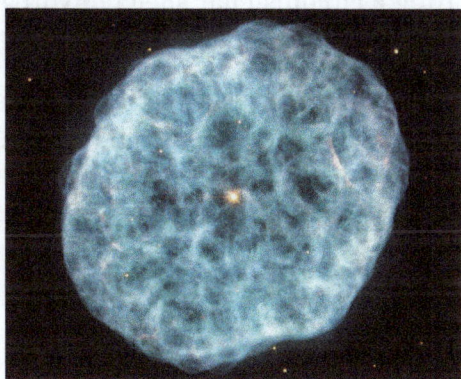

The Oyster Nebula is a planetary nebula located in the constellation of Camelopardalis.

Planetary nebulae are the remnants of the final stages of stellar evolution for lower-mass stars. Evolved asymptotic giant branch stars expel their outer layers outwards due to strong stellar winds, thus forming gaseous shells, while leaving behind the star's core in the form of a white dwarf. The hot white dwarf illuminates the expelled gases producing emission nebulae with spectra similar to those of emission nebulae found in star formation regions. Technically they are H II regions, because most hydrogen are ionized, but are denser and more compact than nebulae found in star formation regions. Planetary nebulae were given their name by the first astronomical observers who were initially unable to distinguish them from planets, and who tended to confuse them with planets, which were of more interest to them. Our Sun is expected to spawn a planetary nebula about 12 billion years after its formation.

## Protoplanetary Nebula

The Westbrook Nebula is an example of a protoplanetary nebula located
in the constellation of Auriga.

A protoplanetary nebula (PPN) is an astronomical object at the short-lived episode during a star's rapid stellar evolution between the late asymptotic giant branch (LAGB) phase and the following planetary nebula (PN) phase. During the AGB phase, the star undergoes mass loss, emitting a circumstellar shell of hydrogen gas. When this phase comes to an end, the star enters the PPN phase.

The PPN is energized by the central star, causing it to emit strong infrared radiation and become a reflection nebula. Collimated stellar winds from the central star shape and shock the shell into an axially symmetric form, while producing a fast moving molecular wind. The exact point when a PPN becomes a planetary nebula (PN) is defined by the temperature of the central star. The PPN phase continues until the central star reaches a temperature of 30,000 K, after which it is hot enough to ionize the surrounding gas.

### Supernova Remnants

The Crab Nebula, an example of a supernova remnant.

A supernova occurs when a high-mass star reaches the end of its life. When nuclear fusion in the core of the star stops, the star collapses. The gas falling inward either rebounds or gets so strongly heated that it expands outwards from the core, thus causing the star to explode. The expanding shell of gas forms a supernova remnant, a special diffuse nebula. Although much of the optical and X-ray emission from supernova remnants originates from ionized gas, a great amount of the radio emission is a form of non-thermal emission called synchrotron emission. This emission originates from high-velocity electrons oscillating within magnetic fields.

## Interstellar Medium

In astronomy, the interstellar medium (ISM) is the matter and radiation that exists in the space between the star systems in a galaxy. This matter includes gas in ionic, atomic, and molecular form, as well as dust and cosmic rays. It fills interstellar space and blends smoothly into the surrounding intergalactic space. The energy that occupies the same volume, in the form of electromagnetic radiation, is the interstellar radiation field.

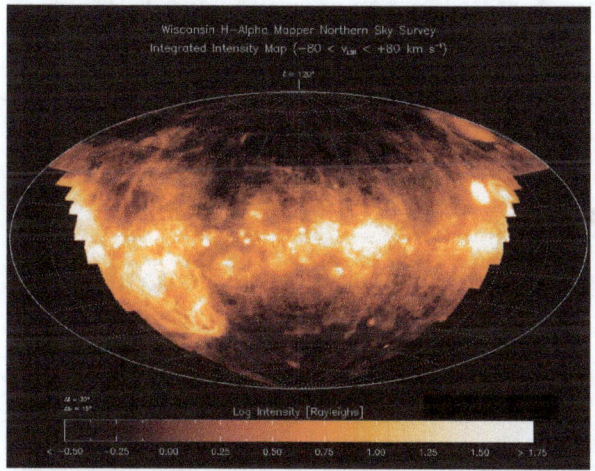

The distribution of ionized hydrogen (known by astronomers as H II from old spectroscopic terminology) in the parts of the Galactic interstellar medium visible from the Earth's northern hemisphere as observed with the Wisconsin Hα Mapper.

The interstellar medium is composed of multiple phases, distinguished by whether matter is ionic, atomic, or molecular, and the temperature and density of the matter. The interstellar medium is composed primarily of hydrogen followed by helium with trace amounts of carbon, oxygen, and nitrogen comparatively to hydrogen. The thermal pressures of these phases are in rough equilibrium with one another. Magnetic fields and turbulent motions also provide pressure in the ISM, and are typically more important dynamically than the thermal pressure is.

In all phases, the interstellar medium is extremely tenuous by terrestrial standards. In cool, dense regions of the ISM, matter is primarily in molecular form, and reaches number densities of $10^6$ molecules per $cm^3$ (1 million molecules per $cm^3$). In hot, diffuse regions of the ISM, matter is primarily ionized, and the density may be as low as $10^{-4}$ ions per $cm^3$. Compare this with a number density of roughly $10^{19}$ molecules per $cm^3$ for air at sea level, and $10^{10}$ molecules per $cm^3$ (10 billion molecules per $cm^3$) for a laboratory high-vacuum chamber. By mass, 99% of the ISM is gas in any form, and 1% is dust. Of the gas in the ISM, by number 91% of atoms are hydrogen and 8.9% are helium, with 0.1% being atoms of elements heavier than hydrogen or helium, known as "metals" in astronomical parlance. By mass this amounts to 70% hydrogen, 28% helium, and 1.5% heavier elements. The hydrogen and helium are primarily a result of primordial nucleosynthesis, while the heavier elements in the ISM are mostly a result of enrichment in the process of stellar evolution.

The ISM plays a crucial role in astrophysics precisely because of its intermediate role between stellar and galactic scales. Stars form within the densest regions of the ISM, which ultimately contributes to molecular clouds and replenishes the ISM with matter and energy through planetary nebulae, stellar winds, and supernovae. This interplay between stars and the ISM helps determine the rate at which a galaxy depletes its gaseous content, and therefore its lifespan of active star formation.

*Voyager 1* reached the ISM on August 25, 2012, making it the first artificial object from Earth to do so. Interstellar plasma and dust will be studied until the mission's end in 2025. Its twin, *Voyager 2* entered the ISM in November 2018.

## Interstellar Matter

Table shows a breakdown of the properties of the components of the ISM of the Milky Way.

Table: Components of the interstellar medium.

| Component | Fractional volume | Scale height (pc) | Temperature (K) | Density (particles/cm³) | State of hydrogen | Primary observational techniques |
|---|---|---|---|---|---|---|
| Molecular clouds | < 1% | 80 | 10–20 | $10^2$–$10^6$ | molecular | Radio and infrared molecular emission and absorption lines |
| Cold Neutral Medium (CNM) | 1–5% | 100–300 | 50–100 | 20–50 | neutral atomic | H I 21 cm line absorption |
| Warm Neutral Medium (WNM) | 10–20% | 300–400 | 6000–10000 | 0.2–0.5 | neutral atomic | H I 21 cm line emission |
| Warm Ionized Medium (WIM) | 20–50% | 1000 | 8000 | 0.2–0.5 | ionized | Hα emission and pulsar dispersion |
| H II regions | < 1% | 70 | 8000 | $10^2$–$10^4$ | ionized | Hα emission and pulsar dispersion |
| Coronal gas Hot Ionized Medium (HIM) | 30–70% | 1000–3000 | $10^6$–$10^7$ | $10^{-4}$–$10^{-2}$ | ionized (metals also highly ionized) | X-ray emission; absorption lines of highly ionized metals, primarily in the ultraviolet |

## The Three-phase Model

Field, Goldsmith & Habing (1969) put forward the static two *phase* equilibrium model to explain the observed properties of the ISM. Their modeled ISM consisted of a cold dense phase ($T < 300$ K), consisting of clouds of neutral and molecular hydrogen, and a warm intercloud phase ($T \sim 10^4$ K), consisting of rarefied neutral and ionized gas. McKee & Ostriker added a dynamic third phase that represented the very hot ($T \sim 10^6$ K) gas which had been shock heated by supernovae and constituted most of the volume of the ISM. These phases are the temperatures where heating and cooling can reach a stable equilibrium. Their paper formed the basis for further study over the past three decades. However, the relative proportions of the phases and their subdivisions are still not well known.

## The Atomic Hydrogen Model

This model takes into account only atomic hydrogen: Temperature larger than 3000 K breaks molecules, lower than 50 000 K leaves atoms in their ground state. It is assumed that influence of other atoms is negligible. Pressure is assumed very low, so that durations of free paths of atoms are larger than the ~ 1 nanosecond duration of light pulses which make ordinary, temporally incoherent light.

In this collisionless gas, Einstein's theory of coherent light-matter interactions applies, all gas-light interactions are spatially coherent. Suppose that a monochromatic light is pulsed, then scattered by molecules having a quadrupole (Raman) resonance frequency. If "length of light pulses is shorter than all involved time constants", an "impulsive stimulated Raman scattering (ISRS) " (Yan, Gamble & Nelson works: While light generated by incoherent Raman at a shifted frequency has a phase independent on phase of exciting light, thus generates a new spectral line, coherence between incident and scattered light allows their interference into a single frequency, thus shifts incident frequency. Assume that a star radiates a continuous light spectrum up to X rays. Lyman frequencies are absorbed in this light and pump atoms mainly to first excited state. In this state, hyperfine periods are longer than 1 ns, so that an ISRS "may" redshift light frequency, populating high hyperfine levels. Another ISRS "may" transfer energy from hyperfine levels to thermal electromagnetic waves, so that redshift is permanent. Temperature of a light beam is defined from frequency and spectral radiance by Planck's formula. As entropy must increase, "may" becomes "does". However, where a previously absorbed line (first Lyman beta) reaches Lyman alpha frequency, redshifting process stops and all hydrogen lines are strongly absorbed. But the stop is not perfect if there is energy at frequency shifted to Lyman beta frequency, which produces a slow redshift. Successive redshifts separated by Lyman absorptions generate many absorption lines, frequencies of which, deduced from absorption process, obey a law more dependable than Karlsson's formula.

The previous process excites more and more atoms because a de-excitation obeys Einstein's law of coherent interactions: Variation $dI$ of radiance $I$ of a light beam along a path $dx$ is $dI=BIdx$, where $B$ is Einstein amplification coefficient which depends on medium. $I$ is the modulus of Poynting vector of field, absorption occurs for an opposed vector, which corresponds to a change of sign of $B$. Factor $I$ in this formula shows that intense rays are more amplified than weak ones (competition of modes). Emission of a flare requires a sufficient radiance $I$ provided by random zero point field. After emission of a flare, weak $B$ increases by pumping while $I$ remains close to zero: De-excitation by a coherent emission involves stochastic parameters of zero point field, as observed close to quasars (and in polar auroras).

## Structures

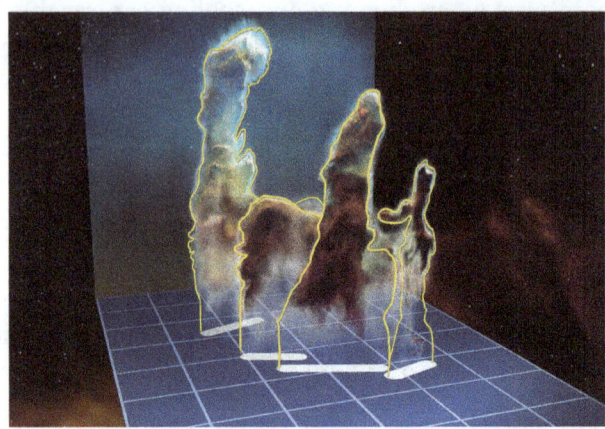

Three-dimensional structure in Pillars of Creation.

The ISM is turbulent and therefore full of structure on all spatial scales. Stars are born deep inside large complexes of molecular clouds, typically a few parsecs in size. During their lives and deaths, stars interact physically with the ISM.

Stellar winds from young clusters of stars (often with giant or supergiant HII regions surrounding them) and shock waves created by supernovae inject enormous amounts of energy into their surroundings, which leads to hypersonic turbulence. The resultant structures – of varying sizes – can be observed, such as stellar wind bubbles and superbubbles of hot gas, seen by X-ray satellite telescopes or turbulent flows observed in radio telescope maps.

The Sun is currently traveling through the Local Interstellar Cloud, a denser region in the low-density Local Bubble.

## Interaction with Interplanetary Medium

The interstellar medium begins where the interplanetary medium of the Solar System ends. The solar wind slows to subsonic velocities at the termination shock, 90–100 astronomical units from the Sun. In the region beyond the termination shock, called the heliosheath, interstellar matter interacts with the solar wind. Voyager 1, the farthest human-made object from the Earth (after 1998 ), crossed the termination shock December 16, 2004 and later entered interstellar space when it crossed the heliopause on August 25, 2012, providing the first direct probe of conditions in the ISM.

## Interstellar Extinction

The ISM is also responsible for extinction and reddening, the decreasing light intensity and shift in the dominant observable wavelengths of light from a star. These effects are caused by scattering and absorption of photons and allow the ISM to be observed with the naked eye in a dark sky. The apparent rifts that can be seen in the band of the Milky

Way – a uniform disk of stars – are caused by absorption of background starlight by molecular clouds within a few thousand light years from Earth.

Far ultraviolet light is absorbed effectively by the neutral components of the ISM. For example, a typical absorption wavelength of atomic hydrogen lies at about 121.5 nanometers, the Lyman-alpha transition. Therefore, it is nearly impossible to see light emitted at that wavelength from a star farther than a few hundred light years from Earth, because most of it is absorbed during the trip to Earth by intervening neutral hydrogen.

## Heating and Cooling

The ISM is usually far from thermodynamic equilibrium. Collisions establish a Maxwell–Boltzmann distribution of velocities, and the 'temperature' normally used to describe interstellar gas is the 'kinetic temperature', which describes the temperature at which the particles would have the observed Maxwell–Boltzmann velocity distribution in thermodynamic equilibrium. However, the interstellar radiation field is typically much weaker than a medium in thermodynamic equilibrium; it is most often roughly that of an A star (surface temperature of ~10,000 K) highly diluted. Therefore, bound levels within an atom or molecule in the ISM are rarely populated according to the Boltzmann formula.

Depending on the temperature, density, and ionization state of a portion of the ISM, different heating and cooling mechanisms determine the temperature of the gas.

## Heating Mechanisms

- Heating by low-energy cosmic rays:

  The first mechanism proposed for heating the ISM was heating by low-energy cosmic rays. Cosmic rays are an efficient heating source able to penetrate in the depths of molecular clouds. Cosmic rays transfer energy to gas through both ionization and excitation and to free electrons through Coulomb interactions. Low-energy cosmic rays (a few MeV) are more important because they are far more numerous than high-energy cosmic rays.

- Photoelectric heating by grains:

  The ultraviolet radiation emitted by hot stars can remove electrons from dust grains. The photon is absorbed by the dust grain, and some of its energy is used to overcome the potential energy barrier and remove the electron from the grain. This potential barrier is due to the binding energy of the electron (the work function) and the charge of the grain. The remainder of the photon's energy gives the ejected electron kinetic energy which heats the gas through collisions with other particles. A typical size distribution of dust grains is $n(r) \propto r^{-3.5}$, where $r$ is the radius of the dust particle. Assuming this, the projected grain surface area distribution is $\pi r^2 n(r) \propto r^{-1.5}$. This indicates that the smallest dust grains dominate this method of heating.

- Photoionization:

When an electron is freed from an atom (typically from absorption of a UV photon) it carries kinetic energy away of the order $E_{photon} - E_{ionization}$. This heating mechanism dominates in H II regions, but is negligible in the diffuse ISM due to the relative lack of neutral carbon atoms.

- X-ray heating:

X-rays remove electrons from atoms and ions, and those photoelectrons can provoke secondary ionizations. As the intensity is often low, this heating is only efficient in warm, less dense atomic medium (as the column density is small). For example, in molecular clouds only hard x-rays can penetrate and x-ray heating can be ignored. This is assuming the region is not near an x-ray source such as a supernova remnant.

- Chemical heating:

Molecular hydrogen ($H_2$) can be formed on the surface of dust grains when two H atoms (which can travel over the grain) meet. This process yields 4.48 eV of energy distributed over the rotational and vibrational modes, kinetic energy of the $H_2$ molecule, as well as heating the dust grain. This kinetic energy, as well as the energy transferred from de-excitation of the hydrogen molecule through collisions, heats the gas.

- Grain-gas heating:

Collisions at high densities between gas atoms and molecules with dust grains can transfer thermal energy. This is not important in HII regions because UV radiation is more important. It is also not important in diffuse ionized medium due to the low density. In the neutral diffuse medium grains are always colder, but do not effectively cool the gas due to the low densities.

Grain heating by thermal exchange is very important in supernova remnants where densities and temperatures are very high.

Gas heating via grain-gas collisions is dominant deep in giant molecular clouds (especially at high densities). Far infrared radiation penetrates deeply due to the low optical depth. Dust grains are heated via this radiation and can transfer thermal energy during collisions with the gas. A measure of efficiency in the heating is given by the accommodation coefficient:

$$\alpha = \frac{T_2 - T}{T_d - T}$$

where $T$ is the gas temperature, $T_d$ the dust temperature, and $T_2$ the post-collision temperature of the gas atom or molecule. This coefficient was measured by (Burke & Hollenbach 1983) as $\alpha = 0.35$.

## Other Heating Mechanisms

A variety of macroscopic heating mechanisms are present including:

- Gravitational collapse of a cloud.

- Supernova explosions.

- Stellar winds.

- Expansion of H II regions.

- Magnetohydrodynamic waves created by supernova remnants.

## Cooling Mechanisms

- Fine structure cooling:

  The process of fine structure cooling is dominant in most regions of the Interstellar Medium, except regions of hot gas and regions deep in molecular clouds. It occurs most efficiently with abundant atoms having fine structure levels close to the fundamental level such as: C II and O I in the neutral medium and O II, O III, N II, N III, Ne II and Ne III in H II regions. Collisions will excite these atoms to higher levels, and they will eventually de-excite through photon emission, which will carry the energy out of the region.

- Cooling by Permitted lines:

  At lower temperatures, more levels than fine structure levels can be populated via collisions. For example, collisional excitation of the $n = 2$ level of hydrogen will release a Ly-$\alpha$ photon upon de-excitation. In molecular clouds, excitation of rotational lines of CO is important. Once a molecule is excited, it eventually returns to a lower energy state, emitting a photon which can leave the region, cooling the cloud.

## Radiowave Propagation

Radio waves from $\approx 10$ kHz (very low frequency) to $\approx 300$ GHz (extremely high frequency) propagate differently in interstellar space than on the Earth's surface. There are many sources of interference and signal distortion that do not exist on Earth. A great deal of radio astronomy depends on compensating for the different propagation effects to uncover the desired signal.

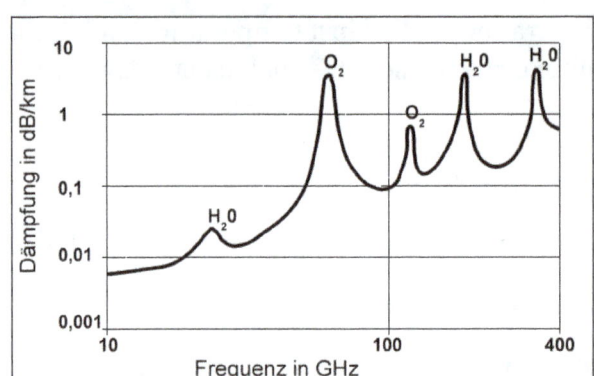

Atmospheric attenuation in dB/km as a function of frequency over the EHF band. Peaks in absorption at specific frequencies are a problem, due to atmosphere constituents such as water vapor ($H_2O$) and carbon dioxide ($CO_2$).

## Knowledge of Interstellar Space

Herbig–Haro 110 object ejects gas through interstellar space.

The nature of the interstellar medium has received the attention of astronomers and scientists over the centuries, and understanding of the ISM has developed. However, they first had to acknowledge the basic concept of "interstellar" space. The term appears to have been first used in print by Bacon (1626, § 354–5): "The Interstellar Skie.. hath so much Affinity with the Starre, that there is a Rotation of that, as well as of the Starre." Later, natural philosopher Robert Boyle (1674) discussed "The inter-stellar part of heaven, which several of the modern Epicureans would have to be empty."

Before modern electromagnetic theory, early physicists postulated that an invisible luminiferous aether existed as a medium to carry lightwaves. It was assumed that this aether extended into interstellar space, as Patterson (1862) wrote, "this efflux occasions a thrill, or vibratory motion, in the ether which fills the interstellar spaces."

The advent of deep photographic imaging allowed Edward Barnard to produce the first images of dark nebulae silhouetted against the background star field of the galaxy, while the first actual detection of cold diffuse matter in interstellar space was made by Johannes Hartmann in 1904 through the use of absorption line spectroscopy. In his historic study of the spectrum and orbit of Delta Orionis, Hartmann observed the light coming from this star and realized that some of this light was being absorbed before it reached the Earth. Hartmann reported that absorption from the "K" line of calcium appeared "extraordinarily weak, but almost perfectly sharp" and also reported the "quite surprising result that the calcium line at 393.4 nanometres does not share in the periodic displacements of the lines caused by the orbital motion of the spectroscopic binary star". The stationary nature of the line led Hartmann to conclude that the gas responsible for the absorption was not present in the atmosphere of Delta Orionis, but was instead located within an isolated cloud of matter residing somewhere along the line-of-sight to this star. This discovery launched the study of the Interstellar Medium.

In the series of investigations, Viktor Ambartsumian introduced the now commonly accepted notion that interstellar matter occurs in the form of clouds.

Following Hartmann's identification of interstellar calcium absorption, interstellar sodium was detected by Heger (1919) through the observation of stationary absorption from the atom's "D" lines at 589.0 and 589.6 nanometres towards Delta Orionis and Beta Scorpii.

Subsequent observations of the "H" and "K" lines of calcium by Beals (1936) revealed double and asymmetric profiles in the spectra of Epsilon and Zeta Orionis. These were the first steps in the study of the very complex interstellar sightline towards Orion. Asymmetric absorption line profiles are the result of the superposition of multiple absorption lines, each corresponding to the same atomic transition (for example the "K" line of calcium), but occurring in interstellar clouds with different radial velocities. Because each cloud has a different velocity (either towards or away from the observer/Earth) the absorption lines occurring within each cloud are either Blue-shifted or Red-shifted (respectively) from the lines' rest wavelength, through the Doppler Effect. These observations confirming that matter is not distributed homogeneously were the first evidence of multiple discrete clouds within the ISM.

The growing evidence for interstellar material led Pickering (1912) to comment that "While the interstellar absorbing medium may be simply the ether, yet the character of its selective absorption, as indicated by Kapteyn, is characteristic of a gas, and free gaseous molecules are certainly there, since they are probably constantly being expelled by the Sun and stars."

The same year Victor Hess's discovery of cosmic rays, highly energetic charged particles that rain onto the Earth from space, led others to speculate whether they also pervaded interstellar space. The following year the Norwegian explorer and physicist Kristian Birkeland wrote: "It seems to be a natural consequence of our points of view to assume that the whole of space is filled with electrons and flying electric ions of all

kinds. We have assumed that each stellar system in evolutions throws off electric cor-
puscles into space. It does not seem unreasonable therefore to think that the greater
part of the material masses in the universe is found, not in the solar systems or nebulae,
but in 'empty' space".

This light-year-long knot of interstellar gas and dust resembles a caterpillar.

Thorndike (1930) noted that "it could scarcely have been believed that the enormous
gaps between the stars are completely void. Terrestrial aurorae are not improbably
excited by charged particles emitted by the Sun. If the millions of other stars are
also ejecting ions, as is undoubtedly true, no absolute vacuum can exist within the
galaxy."

In September 2012, NASA scientists reported that polycyclic aromatic hydrocarbons
(PAHs), subjected to *interstellar medium (ISM)* conditions, are transformed, through
hydrogenation, oxygenation and hydroxylation, to more complex organics – "a step
along the path toward amino acids and nucleotides, the raw materials of proteins and
DNA, respectively". Further, as a result of these transformations, the PAHs lose their
spectroscopic signature which could be one of the reasons "for the lack of PAH detection in interstellar ice grains, particularly the outer regions of cold, dense clouds or the
upper molecular layers of protoplanetary disks."

In February 2014, NASA announced a greatly upgraded database for tracking polycyclic aromatic hydrocarbons (PAHs) in the universe. According to scientists, more
than 20% of the carbon in the universe may be associated with PAHs, possible starting
materials for the formation of life. PAHs seem to have been formed shortly after the
Big Bang, are widespread throughout the universe, and are associated with new stars
and exoplanets.

In April 2019, scientists, working with the Hubble Space Telescope, reported the confirmed detection of the large and complex ionized molecules of buckminsterfuller-
ene ($C_{60}$) (also known as "buckyballs") in the interstellar medium spaces between the
stars.

# References

- Schmadel, Lutz D.; International Astronomical Union (2003). Dictionary of minor planet names. Berlin; New York: Springer-Verlag. pp. 592–593. ISBN 978-3-540-00238-3. Retrieved 9 September 2011

- Celestial-bodies, physics: byjus.com, Retrieved 31 July, 2019

- "The Pillars of Creation Revealed in 3D". European Southern Observatory. 30 April 2015. Retrieved 14 June 2015

- Star-astronomy, science: britannica.com, Retrieved 25 August, 2019

- Cordiner, M.A.; et al. (22 April 2019). "Confirming Interstellar C60 + Using the Hubble Space Telescope". The Astrophysical Journal Letters. 875 (2): L28. doi:10.3847/2041-8213/ab14e5

- What-is-a-galaxy, galaxies, the-universe, partner-content: khanacademy.org, Retrieved 12 May, 2019

# Black Hole | 5

- **Event Horizon**
- **Gravitational Singularity**
- **Photon Sphere**
- **Ergosphere**
- **Hawking Radiation**

Black hole is a region of space-time that has extreme gravitational acceleration due to which no particles or electromagnetic radiation can escape from it. The chapter closely examines the properties and structure of black hole such as event horizon, ergosphere and photon sphere to provide an extensive understanding of the subject.

Black holes are volumes of space where gravity is extreme enough to prevent the escape of even the fastest moving particles. Not even light can break free, hence the name 'black' hole.

A German physicist and astronomer named Karl Schwarzschild proposed the modern version of a black hole in 1915 after coming up with an exact solution to Einstein's approximations of general relativity.

Schwarzschild realised it was possible for mass to be squeezed into an infinitely small point. This would make spacetime around it bend so that nothing – not even massless photons of light – could escape its curvature.

The cusp of the black hole's slide into oblivion is today referred to as its event horizon, and the distance between this boundary and the infinitely dense core – or singularity – is named after Schwarzschild.

Theoretically, all masses have a Schwarzschild radius that can be calculated. If the Sun's mass was squeezed into an infinitely small point, it would form a black hole with a radius of just under 3 kilometres (about 2 miles).

Similarly, Earth's mass would have a Schwarzschild radius of just a few millimetres, making a black hole no bigger than a marble.

For decades, black holes were exotic peculiarities of general relativity. Physicists have became increasingly confident in their existence as other extreme astronomical objects, such as neutron stars, were discovered. Today it's believed most galaxies have monstrous black holes at their core.

## Formation of Black Holes

It's generally accepted that stars with a mass at least three times greater than that of our Sun's can undergo extreme gravitational collapse once their fuel depletes.

With so much mass in a confined volume, the collective force of gravity overcomes the rule that usually keeps the building blocks of atoms from occupying the same space. All this density creates a black hole.

A second type of miniature black hole has been hypothesised, though never observed. They're thought to have formed when the rippling vacuum of the early Universe rapidly expanded in an event known as inflation, causing highly dense regions to collapse.

## Event Horizon

A black hole's mass is concentrated at a single point deep in its heart, and clearly cannot be seen. The "hole" that can, in principle, be seen (although no-one has ever actually seen a black hole directly) is the region of space around the singularity where gravity is so strong that nothing, not even light, the fastest thing in the universe, can escape, and where the time dilation becomes almost infinite.

A black hole is therefore bounded by a well-defined surface or edge known as the "event horizon", within which nothing can be seen and nothing can escape, because the necessary escape velocity would equal or exceed the speed of light (a physical impossibility). The event horizon acts like a kind of one-way membrane, similar to the "point-of-no-return" a boat experiences when approaching a whirlpool and reaching the point where it is no longer possible to navigate against the flow. Or, to look at it in a different way, within the event horizon, space itself is falling into the black hole at a notional speed greater than the speed of light.

The event horizon of a black hole from an exploding star with a mass of several times that of our own Sun, would be perhaps a few kilometers across. However, it could then grow over time as it swallowed dust, planets, stars, even other black holes. The black hole at the center of the Milky Way, for example, is estimated to have a mass equal to about 2,500,000 suns and have an event horizon many millions of kilometers across.

Material, such as gas, dust and other stellar debris that has come close to a black hole but not quite fallen into it, forms a flattened band of spinning matter around the event horizon called the accretion disk (or disc). Although no-one has ever actually seen a black hole or even its event horizon, this accretion disk can be seen, because the spinning particles are accelerated to tremendous speeds by the huge gravity of the black hole, releasing heat and powerful x-rays and gamma rays out into the universe as they smash into each other.

These accretion disks are also known as quasars (quasi-stellar radio sources). Quasars are the oldest known bodies in the universe and (with the exception of gamma ray bursts) the most distant objects we can actually see, as well as being the brightest and most massive, outshining trillions of stars. A quasar is, then, a bright halo of matter surrounding, and being drawn into, a rotating black hole, effectively feeding it with matter. A quasar dims into a normal black hole when there is no matter around it left to eat.

A non-rotating black hole would be precisely spherical. However, a rotating black hole (created from the collapse of a rotating star) bulges out at its equator due to centripetal force. A rotating black hole is also surrounded by a region of space-time in which it is impossible to stand still, called the ergosphere. This is due to a process known as frame-dragging, whereby any rotating mass will tend to slightly "drag" along the space-time immediately surrounding it. In fact, space-time in the ergosphere is technically dragged around faster than the speed of light (relative, that is, to other regions of space-time surrounding it). It may be possible for objects in the ergosphere to escape from orbit around the black hole but, once within the ergosphere, they cannot remain stationary.

Also due to the extreme gravity around a black hole, an object in its gravitational field experiences a slowing down of time, known as gravitational time dilation, relative to observers outside the field. From the viewpoint of a distant observer an object falling into a black hole appears to slow down and fade, approaching but never quite reaching the event horizon. Finally, at a point just before it reaches the event horizon, it becomes so dim that it can no longer be seen (all due to the time dilation effect).

## Gravitational Singularity

A gravitational singularity (or space-time singularity) is a location where the quantities that are used to measure the gravitational field become infinite in a way that does not depend on the coordinate system. In other words, it is a point in which all physical laws are indistinguishable from one another, where space and time are no longer interrelated realities, but merge indistinguishably and cease to have any independent meaning.

This artist's impression depicts a rapidly spinning supermassive
black hole surrounded by an accretion disc.

## Origin of Theory

Singularities were first predicated as a result of Einstein's Theory of General Relativity, which resulted in the theoretical existence of black holes. In essence, the theory predicted that any star reaching beyond a certain point in its mass (aka. the Schwarzschild Radius) would exert a gravitational force so intense that it would collapse.

At this point, nothing would be capable of escaping its surface, including light. This is due to the fact the gravitational force would exceed the speed of light in vacuum – 299,792,458 meters per second (1,079,252,848.8 km/h; 670,616,629 mph).

This phenomena is known as the Chandrasekhar Limit, named after the Indian astrophysicist Subrahmanyan Chandrasekhar, who proposed it in 1930. At present, the accepted value of this limit is believed to be 1.39 Solar Masses (i.e. 1.39 times the mass of our Sun), which works out to a whopping $2.765 \times 10^{30}$ kg (or 2,765 trillion trillion metric tons).

Another aspect of modern General Relativity is that at the time of the Big Bang (i.e. the initial state of the Universe) was a singularity. Roger Penrose and Stephen Hawking both developed theories that attempted to answer how gravitation could produce singularities, which eventually merged together to be known as the Penrose–Hawking Singularity Theorems.

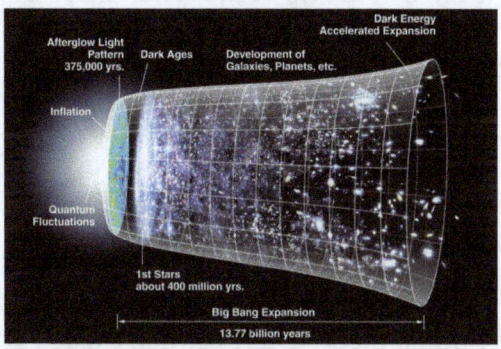

The Big Bang Theory: A history of the Universe starting from a singularity
and expanding ever since.

According to the Penrose Singularity Theorem, which he proposed in 1965, a time-like singularity will occur within a black hole whenever matter reaches certain energy conditions. At this point, the curvature of space-time within the black hole becomes infinite, thus turning it into a trapped surface where time ceases to function.

The Hawking Singularity Theorem added to this by stating that a space-like singularity can occur when matter is forcibly compressed to a point, causing the rules that govern matter to break down. Hawking traced this back in time to the Big Bang, which he claimed was a point of infinite density. However, Hawking later revised this to claim that general relativity breaks down at times prior to the Big Bang, and hence no singularity could be predicted by it.

Some more recent proposals also suggest that the Universe did not begin as a singularity. These includes theories like Loop Quantum Gravity, which attempts to unify the laws of quantum physics with gravity. This theory states that, due to quantum gravity effects, there is a minimum distance beyond which gravity no longer continues to increase, or that interpenetrating particle waves mask gravitational effects that would be felt at a distance.

## Types of Singularities

The two most important types of space-time singularities are known as Curvature Singularities and Conical Singularities. Singularities can also be divided according to whether they are covered by an event horizon or not. In the case of the former, you have the Curvature and Conical; whereas in the latter, you have what are known as Naked Singularities.

A Curvature Singularity is best exemplified by a black hole. At the center of a black hole, space-time becomes a one-dimensional point which contains a huge mass. As a result, gravity become infinite and space-time curves infinitely, and the laws of physics as we know them cease to function.

Conical singularities occur when there is a point where the limit of every general covariance quantity is finite. In this case, space-time looks like a cone around this point, where the singularity is located at the tip of the cone. An example of such a conical singularity is a cosmic string, a type of hypothetical one-dimensional point that is believed to have formed during the early Universe.

And, as mentioned, there is the Naked Singularity, a type of singularity which is not hidden behind an event horizon. These were first discovered in 1991 by Shapiro and Teukolsky using computer simulations of a rotating plane of dust that indicated that General Relativity might allow for "naked" singularities.

In this case, what actually transpires within a black hole (i.e. its singularity) would be visible. Such a singularity would theoretically be what existed prior to the Big Bang. The key word here is theoretical, as it remains a mystery what these objects would look like.

For the moment, singularities and what actually lies beneath the veil of a black hole remains a mystery. As time goes on, it is hoped that astronomers will be able to study black holes in greater detail. It is also hoped that in the coming decades, scientists will find a way to merge the principles of quantum mechanics with gravity, and that this will shed further light on how this mysterious force operates.

## Photon Sphere

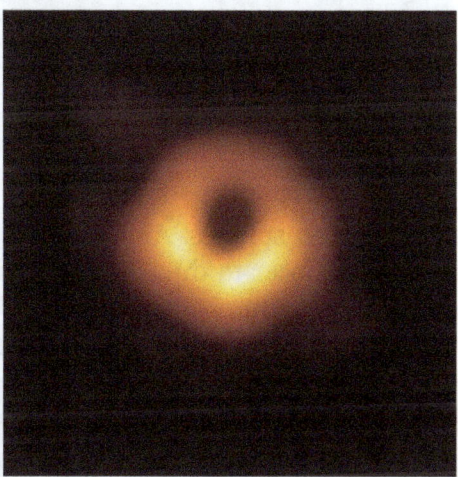

Radio emission from the accretion disk surrounding the supermassive black hole M87* (captured 2017, computed 2019) as imaged by the Event Horizon Telescope. The Photon sphere lies within the dark shadow (which has a radius of 2.6 times the Schwarzschild radius).

A photon sphere or photon circle is an area or region of space where gravity is so strong that photons are forced to travel in orbits. (It is sometimes called the last photon orbit.) The radius of the photon sphere, which is also the lower bound for any stable orbit, is, for a Schwarzschild black hole:

$$r = \frac{3GM}{c^2} = \frac{3r_s}{2}$$

where $G$ is the gravitational constant, $M$ is the black hole mass, and $c$ is the speed of light in vacuum and $r_s$ is the Schwarzschild radius (the radius of the event horizon) -

This equation entails that photon spheres can only exist in the space surrounding an extremely compact object (a black hole or possibly an "ultracompact" neutron star ).

The photon sphere is located farther from the center of a black hole than the event horizon. Within a photon sphere, it is possible to imagine a photon that begins at the back of your head, orbiting the black hole, only then to be intercepted by your eyes, allowing you to see the back of your head. For non-rotating black holes, the photon sphere is a sphere of radius 3/2 $r_s$. There are no stable free fall orbits that exist within or cross the photon sphere. Any free fall orbit that crosses it from the outside spirals into the black hole. Any orbit that crosses it from the inside escapes to infinity or falls back in and spirals into the black hole. No unaccelerated orbit with a semi-major axis less than this distance is possible, but within the photon sphere, a constant acceleration will allow a spacecraft or probe to hover above the event horizon.

Another property of the photon sphere is centrifugal force (nb: not centripetal) reversal. Outside the photon sphere, the faster one orbits the greater the outward force one feels. Centrifugal force falls to zero at the photon sphere, including non-freefall orbits at any speed, i.e. you weigh the same no matter how fast you orbit, and becomes negative inside it. Inside the photon sphere the faster you orbit the greater your felt weight or inward force. This has serious ramifications for the fluid dynamics of inward fluid flow.

A rotating black hole has two photon spheres. As a black hole rotates, it drags space with it. The photon sphere that is closer to the black hole is moving in the same direction as the rotation, whereas the photon sphere further away is moving against it. The greater the angular velocity of the rotation of a black hole, the greater the distance between the two photon spheres. Since the black hole has an axis of rotation, this only holds true if approaching the black hole in the direction of the equator. If approaching at a different angle, such as one from the poles of the black hole to the equator, there is only one photon sphere. This is because approaching at this angle the possibility of traveling with or against the rotation does not exist.

## Derivation for a Schwarzschild Black Hole

Since a Schwarzschild black hole has spherical symmetry, all possible axes for a circular photon orbit are equivalent, and all circular orbits have the same radius.

This derivation involves using the Schwarzschild metric, given by:

$$ds^2 = \left(1 - \frac{r_s}{r}\right)c^2 dt^2 - \left(1 - \frac{r_s}{r}\right)^{-1} dr^2 - r^2(\sin^2\theta d\phi^2 + d\theta^2)$$

For a photon traveling at a constant radius r (i.e. in the $\Phi$-coordinate direction), $dr = 0$. Since it is a photon $ds = 0$ (a "light-like interval"). We can always rotate the coordinate system such that $\theta$ is constant, $d\theta = 0$.

Setting ds, dr and dθ to zero, we have:

$$\left(1 - \frac{r_s}{r}\right)c^2 dt^2 = r^2 \sin^2\theta d\phi^2$$

Re-arranging gives:

$$\frac{\phi}{dt} = \frac{c}{r\sin\theta}\sqrt{1-\frac{r_s}{r}}$$

To proceed we need the relation $\frac{d\phi}{dt}$. To find it, we use the radial geodesic equation,

$$\frac{d^2r}{d\tau^2} + \Gamma^r_{\mu\nu}u^\mu u^\nu = 0$$

Non vanishing $\Gamma$-connection coefficients are:

$$\Gamma^r_{tt} = \frac{c^2 BB'}{2}, \Gamma^r_{rr} = -\frac{B^{-1}B'}{2}, \Gamma^r_{\theta\theta} = -rB, \Gamma^r_{\phi\phi} = -Br\sin^2\theta,$$

where,

$$B' = \frac{dB}{dr}, B = 1-\frac{r_s}{r}.$$

We treat photon radial geodesics with constant r and $\theta$, therefore:

$$\frac{dr}{d\tau}, \frac{d^2r}{d\tau^2}, \frac{d\theta}{d\tau} = 0.$$

Substituting it all into the radial geodesic equation (the geodesic equation with the radial coordinate as the dependent variable), we obtain:

$$\left(\frac{d\phi}{dt}\right)^2 = \frac{c^2 r_s}{2r^3 \sin^2\theta}$$

Comparing it with obtained previously, we have:

$$c\sqrt{\frac{r_s}{2r}} = c\sqrt{1-\frac{r_s}{r}}$$

where we have inserted $\theta = \frac{\pi}{2}$ radians (imagine that the central mass, about which the photon is orbiting, is located at the centre of the coordinate axes. Then, as the photon is travelling along the $\phi$-coordinate line, for the mass to be located directly in the centre of the photon's orbit, we must have $\theta = \frac{\pi}{2}$ radians).

Hence, rearranging this final expression gives:

$$r = \frac{3}{2}r_s$$

which is the result we set out to prove.

## Photon Orbits Around a Kerr Black Hole

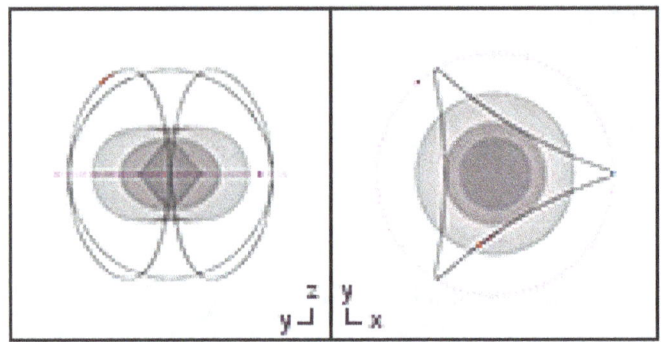

Views from the side (l) and from above a pole (r). A rotating black hole
has 2 radii between which light can orbit on a constant r-coordinate.

In contrast to a Schwarzschild black hole, a Kerr (spinning) black hole does not have spherical symmetry, but only an axis of symmetry, which has profound consequences for the photon orbits, A circular orbit can only exist in the equatorial plane, and there are two of them (prograde and retrograde), with different Boyer–Lindquist-radii,

$$r_{\pm}^{\circ} = r_{s}\left[1 + \cos\left(\frac{2}{3}\cos^{-1}\left(\frac{\pm|a|}{M}\right)\right)\right]$$

where $a$ is the angular momentum. There exist other constant coordinate-radius orbits, but they have more complicated paths which oscillate in latitude about the equator.

## Ergosphere

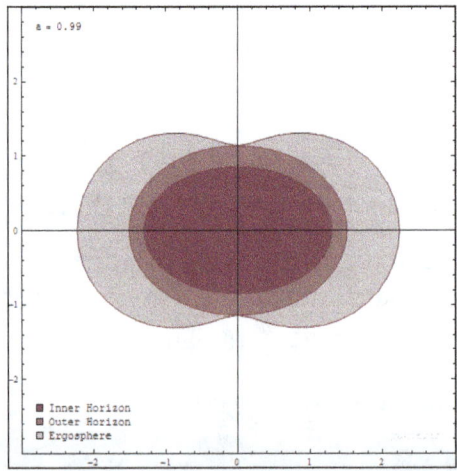

In the ergosphere (shown here in light gray), the component $g_{tt}$ is negative, i.e., acts like a purely spatial metric component. Consequently, timelike or lightlike worldlines within this region must co-rotate with the inner mass. Cartesian Kerr–Schild coordinates, equatorial perspective.

The ergosphere is a region located outside a rotating black hole's outer event horizon. It received this name because it is theoretically possible to extract energy and mass from this region. The ergosphere touches the event horizon at the poles of a rotating black hole and extends to a greater radius at the equator. With a low spin of the central mass the shape of the ergosphere can be approximated by an oblate spheroid, while with higher spins it resembles a pumpkin shape. The equatorial (maximal) radius of an ergosphere corresponds to the Schwarzschild radius of a non-rotating black hole; the polar (minimal) radius can be as little as half the Schwarzschild radius (the radius of a non-rotating black hole) in the case that the black hole is rotating maximally (at higher rotation rates the black hole could not have formed).

## Rotation

As a black hole rotates, it twists spacetime in the direction of the rotation at a speed that decreases with distance from the event horizon. This process is known as the Lense–Thirring effect or frame-dragging. Because of this dragging effect, an object within the ergosphere cannot appear stationary with respect to an outside observer at a great distance unless that object were to move at faster than the speed of light (an impossibility) with respect to the local spacetime. The speed necessary for such an object to appear stationary decreases at points further out from the event horizon, until at some distance the required speed is that of the speed of light.

The set of all such points defines the ergosphere surface, called ergosurface. The outer surface of the ergosphere is called the *static surface* or *static limit*. This is because world lines change from being time-like outside the static limit to being space-like inside it. It is the speed of light that arbitrarily defines the ergosphere surface. Such a surface would appear as an oblate that is coincident with the event horizon at the pole of rotation, but at a greater distance from the event horizon at the equator. Outside this surface, space is still dragged, but at a lesser rate.

## Radial Pull

A suspended plumb, held stationary outside the ergosphere, will experience an infinite/diverging radial pull as it approaches the static limit. At some point it will start to fall, resulting in a gravitomagnetically induced spinward motion. An implication of this dragging of space is the existence of negative energies within the ergosphere.

Since the ergosphere is outside the event horizon, it is still possible for objects that enter that region with sufficient velocity to escape from the gravitational pull of the black hole. An object can gain energy by entering the black hole's rotation and then escaping from it, thus taking some of the black hole's energy with it (making the maneuver similar to the exploitation of the Oberth effect around "normal" space objects).

This process of removing energy from a rotating black hole was proposed by the mathematician Roger Penrose in 1969 and is called the Penrose process. The maximal amount of energy gain possible for a single particle via this process is 20.7% in terms of its mass equivalence, and if this process is repeated by the same mass, the theoretical maximal energy gain approaches 29% of its original mass-energy equivalent. As this energy is removed, the black hole loses angular momentum, the limit of zero rotation is approached as spacetime dragging is reduced. In the limit, the ergosphere no longer exists. This process is considered a possible explanation for a source of energy of such energetic phenomena as gamma ray bursts. Results from computer models show that the Penrose process is capable of producing the high-energy particles that are observed being emitted from quasars and other active galactic nuclei.

### Ergosphere Size

The size of the ergosphere, the distance between the ergosurface and the event horizon, is not necessarily proportional to the radius of the event horizon, but rather to the black hole's gravity and its angular momentum. A point at the poles does not move, and thus has no angular momentum, while at the equator a point would have its greatest angular momentum. This variation of angular momentum that extends from the poles to the equator is what gives the ergosphere its oblate shape. As the mass of the black hole or its rotation speed increases, the size of the ergosphere increases as well.

## Hawking Radiation

Simulated view of a black hole (center) in front of the Large Magellanic Cloud. Note the gravitational lensing effect, which produces two enlarged but highly distorted views of the Cloud. Across the top, the Milky Way disk appears distorted into an arc.

Hawking radiation is black-body radiation that is predicted to be released by black holes, due to quantum effects near the event horizon. It is named after the theoretical physicist Stephen Hawking, who provided a theoretical argument for its existence in 1974.

Hawking radiation reduces the mass and rotation energy of black holes and is therefore also known as black hole evaporation. Because of this, black holes that do not gain mass through other means are expected to shrink and ultimately vanish. Micro black holes are predicted to be larger emitters of radiation than larger black holes and should shrink and dissipate faster.

In June 2008, NASA launched the Fermi space telescope, which is searching for the terminal gamma-ray flashes expected from evaporating primordial black holes. In the event that speculative large extra dimension theories are correct, CERN's Large Hadron Collider may be able to create micro black holes and observe their evaporation. No such micro black hole has ever been observed at CERN.

In September 2010, a signal that is closely related to black hole Hawking radiation was claimed to have been observed in a laboratory experiment involving optical light pulses. However, the results remain unverified and debatable. Other projects have been launched to look for this radiation within the framework of analog gravity.

Black holes are sites of immense gravitational attraction. Classically, the gravitation generated by the gravitational singularity inside a black hole is so powerful that nothing, not even electromagnetic radiation, can escape from the black hole. It is yet unknown how gravity can be incorporated into quantum mechanics. Nevertheless, far from the black hole, the gravitational effects can be weak enough for calculations to be reliably performed in the framework of quantum field theory in curved spacetime. Hawking showed that quantum effects allow black holes to emit exact black-body radiation. The electromagnetic radiation is produced as if emitted by a black body with a temperature inversely proportional to the mass of the black hole.

Physical insight into the process may be gained by imagining that particle–antiparticle radiation is emitted from just beyond the event horizon. This radiation does not come directly from the black hole itself, but rather is a result of virtual particles being "boosted" by the black hole's gravitation into becoming real particles. As the particle–antiparticle pair was produced by the black hole's gravitational energy, the escape of one of the particles lowers the mass of the black hole.

An alternative view of the process is that vacuum fluctuations cause a particle–antiparticle pair to appear close to the event horizon of a black hole. One of the pair falls into the black hole while the other escapes. In order to preserve total energy, the particle that fell into the black hole must have had a negative energy (with respect to an observer far away from the black hole). This causes the black hole to lose mass, and, to an outside observer, it would appear that the black hole has just emitted a particle. In another model, the process is a quantum tunnelling effect, whereby particle–antiparticle pairs will form from the vacuum, and one will tunnel outside the event horizon.

An important difference between the black hole radiation as computed by Hawking and thermal radiation emitted from a black body is that the latter is statistical in nature, and only its average satisfies what is known as Planck's law of black-body radiation, while the former fits the data better. Thus thermal radiation contains information about the body that emitted it, while Hawking radiation seems to contain no such information, and depends only on the mass, angular momentum, and charge of the black hole (the no-hair theorem). This leads to the black hole information paradox.

However, according to the conjectured gauge-gravity duality (also known as the AdS/CFT correspondence), black holes in certain cases (and perhaps in general) are equivalent to solutions of quantum field theory at a non-zero temperature. This means that no information loss is expected in black holes (since the theory permits no such loss) and the radiation emitted by a black hole is probably the usual thermal radiation. If this is correct, then Hawking's original calculation should be corrected, though it is not known how.

A black hole of one solar mass ($M_\odot$) has a temperature of only 60 nanokelvins (60 billionths of a kelvin); in fact, such a black hole would absorb far more cosmic microwave background radiation than it emits. A black hole of $4.5 \times 10^{22}$ kg (about the mass of the Moon, or about 133 μm across) would be in equilibrium at 2.7 K, absorbing as much radiation as it emits. Yet smaller primordial black holes would emit more than they absorb and thereby lose mass.

Hawking's discovery followed a visit to Moscow in 1973 where the Soviet scientists Yakov Zel'dovich and Alexei Starobinsky convinced him that rotating black holes ought to create and emit particles. When Hawking did the calculation, he found to his surprise that even non-rotating black holes produce radiation.

## Trans-Planckian Problem

The trans-Planckian problem is the issue that Hawking's original calculation includes quantum particles where the wavelength becomes shorter than the Planck length near the black hole's horizon. This is due to the peculiar behavior there, where time stops as measured from far away. A particle emitted from a black hole with a finite frequency, if traced back to the horizon, must have had an infinite frequency, and therefore a trans-Planckian wavelength.

The Unruh effect and the Hawking effect both talk about field modes in the superficially stationary space-time that change frequency relative to other coordinates that are regular across the horizon. This is necessarily so, since to stay outside a horizon requires acceleration that constantly Doppler shifts the modes.

An outgoing Hawking radiated photon, if the mode is traced back in time, has a frequency that diverges from that which it has at great distance, as it gets closer to the horizon, which requires the wavelength of the photon to "scrunch up" infinitely at the horizon of the black hole. In a maximally extended external Schwarzschild solution, that photon's frequency stays regular only if the mode is extended back into the past region where no observer can go. That region seems to be unobservable and is physically suspect, so Hawking used a black hole solution without a past region that forms at a finite time in the past. In that case, the source of all the outgoing photons can be identified: a microscopic point right at the moment that the black hole first formed.

The quantum fluctuations at that tiny point, in Hawking's original calculation, contain all the outgoing radiation. The modes that eventually contain the outgoing radiation at long times are redshifted by such a huge amount by their long sojourn next to the event horizon, that they start off as modes with a wavelength much shorter than the Planck length. Since the laws of physics at such short distances are unknown, some find Hawking's original calculation unconvincing.

The trans-Planckian problem is nowadays mostly considered a mathematical artifact of horizon calculations. The same effect occurs for regular matter falling onto a white hole solution. Matter that falls on the white hole accumulates on it, but has no future region into which it can go. Tracing the future of this matter, it is compressed onto the final singular endpoint of the white hole evolution, into a trans-Planckian region. The reason for these types of divergences is that modes that end at the horizon from the point of view of outside coordinates are singular in frequency there. The only way to determine what happens classically is to extend in some other coordinates that cross the horizon.

There exist alternative physical pictures that give the Hawking radiation in which the trans-Planckian problem is addressed. The key point is that similar trans-Planckian problems occur when the modes occupied with Unruh radiation are traced back in time. In the Unruh effect, the magnitude of the temperature can be calculated from ordinary Minkowski field theory, and is not controversial.

## Emission Process

Hawking radiation is required by the Unruh effect and the equivalence principle applied to black hole horizons. Close to the event horizon of a black hole, a local observer must accelerate to keep from falling in. An accelerating observer sees a thermal bath of particles that pop out of the local acceleration horizon, turn around, and free-fall back in. The condition of local thermal equilibrium implies that the consistent extension of this local thermal bath has a finite temperature at infinity, which implies that some of these particles emitted by the horizon are not reabsorbed and become outgoing Hawking radiation.

A Schwarzschild black hole has a metric:

$$(ds)^2 = -\left(1 - \tfrac{2M}{r}\right)(dt)^2 + \frac{1}{\left(1 - \dfrac{2M}{r}\right)}(dr)^2 + r^2(d\Omega)^2 .$$

The black hole is the background spacetime for a quantum field theory.

The field theory is defined by a local path integral, so if the boundary conditions at the horizon are determined, the state of the field outside will be specified. To find the appropriate boundary conditions, consider a stationary observer just outside the horizon at position,

$$r = 2M + \frac{\rho^2}{8M}.$$

The local metric to lowest order is

$$\left(\mathrm{d}s\right)^2 = -\left(\frac{\rho}{4M}\right)^2 \left(\mathrm{d}t\right)^2 + \left(\mathrm{d}\rho\right)^2 + \left(\mathrm{d}X_\perp\right)^2 = -\rho^2\left(\mathrm{d}\tau\right)^2 + \left(\mathrm{d}\rho\right)^2 + \left(\mathrm{d}X_\perp\right)^2,$$

which is Rindler in terms of $\tau = t/4M$. The metric describes a frame that is accelerating to keep from falling into the black hole. The local acceleration, $\alpha = 1/\rho$, diverges as $\rho \to 0$.

The horizon is not a special boundary, and objects can fall in. So the local observer should feel accelerated in ordinary Minkowski space by the principle of equivalence. The near-horizon observer must see the field excited at a local temperature,

$$T = \frac{\alpha}{2\pi} = \frac{1}{2\pi\rho} = \frac{1}{4\pi\sqrt{2M(r-2M)}},$$

which is the Unruh effect.

The gravitational redshift is given by the square root of the time component of the metric. So for the field theory state to consistently extend, there must be a thermal background everywhere with the local temperature redshift-matched to the near horizon temperature:

$$T(r') = \frac{1}{4\pi\sqrt{2M(r-2M)}}\sqrt{\frac{1-\frac{2M}{r}}{1-\frac{2M}{r'}}} = \frac{1}{4\pi\sqrt{2Mr\left(1-\frac{2M}{r'}\right)}}.$$

The inverse temperature redshifted to $r'$ at infinity is,

$$T(\infty) = \frac{1}{4\pi\sqrt{2Mr}}$$

and $r$ is the near-horizon position, near $2M$, so this is really:

$$T(\infty) = \frac{1}{8\pi M}.$$

So a field theory defined on a black hole background is in a thermal state whose temperature at infinity is:

$$T_{\mathrm{H}} = \frac{1}{8\pi M}.$$

This can be expressed in a cleaner way in terms of the surface gravity of the black hole; this is the parameter that determines the acceleration of a near-horizon observer. In natural units ($G = c = \hbar = k_{\mathrm{B}} = 1$), the temperature is,

$$T_{\mathrm{H}} = \frac{\kappa}{2\pi},$$

where $\kappa$ is the surface gravity of the horizon. So a black hole can only be in equilibrium with a gas of radiation at a finite temperature. Since radiation incident on the black hole is absorbed, the black hole must emit an equal amount to maintain detailed balance. The black hole acts as a perfect blackbody radiating at this temperature.

In SI units, the radiation from a Schwarzschild black hole is blackbody radiation with temperature,

$$T = \frac{\hbar c^3}{8\pi G k_{\mathrm{B}} M} \approx 1.227 \times 10^{+23}\,\mathrm{K \cdot kg} \times \frac{1}{M} = 6.169 \times 10^{-8}\,\mathrm{K} \times \frac{M_{\odot}}{M},$$

where $\hbar$ is the reduced Planck constant, $c$ is the speed of light, $k_{\mathrm{B}}$ is the Boltzmann constant, $G$ is the gravitational constant, $M_{\odot}$ is the solar mass, and $M$ is the mass of the black hole.

From the black hole temperature, it is straightforward to calculate the black hole entropy. The change in entropy when a quantity of heat $dQ$ is added is:

$$\mathrm{d}S = \frac{\mathrm{d}Q}{T} = 8\pi M \mathrm{d}Q.$$

The heat energy that enters serves to increase the total mass, so:

$$\mathrm{d}S = 8\pi M \mathrm{d}M = \mathrm{d}\!\left(4\pi M^2\right).$$

The radius of a black hole is twice its mass in natural units, so the entropy of a black hole is proportional to its surface area:

$$S = \pi R^2 = \frac{A}{4}.$$

Assuming that a small black hole has zero entropy, the integration constant is zero.

Forming a black hole is the most efficient way to compress mass into a region, and this entropy is also a bound on the information content of any sphere in space time. The form of the result strongly suggests that the physical description of a gravitating theory can be somehow encoded onto a bounding surface.

In 2019, Biancalana, Robson and Villari from Heriot-Watt University in Edinburgh (UK), showed that Hawking's radiation temperature is a purely topological quantity that can be calculated very simply by computing the Euler characteristics of the black hole spacetime.

## Black Hole Evaporation

When particles escape, the black hole loses a small amount of its energy and therefore some of its mass (mass and energy are related by Einstein's equation $E = mc^2$).

## 1976 Page Numerical Analysis

In 1976 Don Page calculated the power produced, and the time to evaporation, for a nonrotating, non-charged Schwarzschild black hole of mass $M$. The calculations are complicated by the fact that a black hole, being of finite size, is not a perfect black body; the absorption cross section goes down in a complicated, spin-dependent manner as frequency decreases, especially when the wavelength becomes comparable to the size of the event horizon. Note that writing in 1976, Page erroneously postulates that neutrinos have no mass and that only two neutrino flavors exist, and therefore his results of black hole lifetimes do not match the modern results which take into account 3 flavors of neutrinos with nonzero masses.

For a mass much larger than $10^{17}$ grams, Page deduces that electron emission can be ignored, and that black holes of mass $M$ in grams evaporate via massless electron and muon neutrinos, photons, and gravitons in a time $\tau$ of,

$$\tau = 8.66 \times 10^{-27} \left[ \frac{M}{g} \right]^3 \text{ s.}$$

For a mass much smaller than $10^{17}$ g, but much larger than $5 \times 10^{14}$ g, the emission of ultrarelativistic electrons and positrons will accelerate the evaporation, giving a lifetime of,

$$\tau = 4.8 \times 10^{-27} \left[ \frac{M}{g} \right]^3 \text{ s.}$$

If black holes evaporate under Hawking radiation, a solar mass black hole will evaporate over $10^{64}$ years. A supermassive black hole with a mass of $10^{11}$ (100 billion) $M_\odot$ will evaporate in around $2 \times 10^{100}$ years. Some monster black holes in the universe are predicted to continue to grow up to perhaps $10^{14}\ M_\odot$ during the collapse of superclusters of galaxies. Even these would evaporate over a timescale of up to $10^{106}$ years.

## A Crude Analytic Estimate

The power emitted by a black hole in the form of Hawking radiation can easily be estimated for the simplest case of a nonrotating, non-charged Schwarzschild black hole of mass $M$. Combining the formulas for the Schwarzschild radius of the black hole, the Stefan–Boltzmann law of blackbody radiation, the above formula for the temperature of the radiation, and the formula for the surface area of a sphere (the black hole's event horizon), several equations can be derived:

Stefan–Boltzmann constant:

$$\sigma = \frac{\pi^2 k_B^4}{60 \hbar^3 c^2}$$

Schwarzschild radius:

$$r_s = \frac{2GM}{c^2}$$

Acceleration due to gravity at the event horizon:

$$g = \frac{GM}{r_s^2} = \frac{c^4}{4GM}$$

Hawking radiation has a blackbody (Planck) spectrum with a temperature $T$ given by:

$$E = k_B T = \frac{\hbar g}{2\pi c} = \frac{\hbar}{2\pi c}\left(\frac{c^4}{4GM}\right) = \frac{\hbar c^3}{8\pi GM}$$

Hawking radiation temperature:

$$T_H = \frac{\hbar c^3}{8\pi GM k_B}$$

For a one solar mass black hole, the peak Hawking radiation temperature is:

$$T_H = \frac{\hbar c^3}{8\pi GM_\odot k_B} = 6.170 \times 10^{-8}\ \text{K}.$$

The peak wavelength of this radiation is nearly 16 times the Schwarzschild radius of the black hole. Using Wien's displacement constant $b = hc/4.9651 k_B = 2.8978 \times 10^{-3}$ m K.

$$\lambda_{max} = \frac{b}{T_H} = \frac{8\pi^2}{4.9651} r_s = 15.902 r_s$$

Schwarzschild sphere surface area of Schwarzschild radius $r_s$:

$$A_s = 4\pi r_s^2 = 4\pi \left(\frac{2GM}{c^2}\right)^2 = \frac{16\pi G^2 M^2}{c^4}$$

Stefan–Boltzmann power law:

$$P = A_s j^* = A_s \varepsilon \sigma T^4$$

For simplicity, assume a black hole is a perfect blackbody ($\varepsilon = 1$).

Stefan–Boltzmann–Schwarzschild–Hawking black hole radiation power law derivation:

$$P = A_s \varepsilon \sigma T_H^4 = \left(\frac{16\pi G^2 M^2}{c^4}\right)\left(\frac{\pi^2 k_B^4}{60\hbar^3 c^2}\right)\left(\frac{\hbar c^3}{8\pi GM k_B}\right)^4 = \frac{\hbar c^6}{15360\pi G^2 M^2}$$

This yields the Bekenstein–Hawking luminosity of a black hole, under the assumption of pure photon emission (no other particles are emitted) and under the assumption that the horizon is the radiating surface:

$$P = \frac{\hbar c^6}{15360\pi G^2 M^2}$$

where $P$ is the luminosity, i.e., the radiated power, $\hbar$ is the reduced Planck constant, $c$ is the speed of light, $G$ is the gravitational constant and $M$ is the mass of the black hole. It is worth mentioning that the above formula has not yet been derived in the framework of semiclassical gravity.

Substituting the numerical values of the physical constants in the formula for luminosity we obtain P= 3.562×10^32 W kg^2/M^2. The power of the Hawking radiation from a solar mass ($M_\odot$) black hole turns out to be minuscule:

$$P = \frac{\hbar c^6}{15360\pi G^2 M_\odot^2} = 9.007 \times 10^{-29} \text{ W}.$$

It is indeed an extremely good approximation to call such an object 'black'. Under the assumption of an otherwise empty universe, so that no matter, cosmic microwave background radiation, or other radiation falls into the black hole, it is possible to calculate how long it would take for the black hole to dissipate:

$$K_{ev} = \frac{\hbar c^6}{15360\pi G^2} = 3.562 \times 10^{32} \text{ W kg}^2.$$

Given that the power of the Hawking radiation is the rate of evaporation energy loss of the black hole:

$$P = -\frac{dE}{dt} = \frac{K_{ev}}{M^2}.$$

Since the total energy $E$ of the black hole is related to its mass $M$ by Einstein's mass–energy formula $E = Mc^2$:

$$P = -\frac{dE}{dt} = -\left(\frac{d}{dt}\right)Mc^2 = -c^2\frac{dM}{dt}.$$

We can then equate this to our above expression for the power:

$$-c^2\frac{dM}{dt} = \frac{K_{ev}}{M^2}.$$

This differential equation is separable, and we can write:

$$M^2\,dM = -\frac{K_{ev}}{c^2}dt.$$

The black hole's mass is now a function $M(t)$ of time $t$. Integrating over $M$ from $M_0$ (the initial mass of the black hole) to zero (complete evaporation), and over $t$ from zero to $t_{ev}$:

$$\int_{M_0}^{0} M^2\,dM = -\frac{K_{ev}}{c^2}\int_0^{t_{ev}} dt$$

The evaporation time of a black hole is proportional to the cube of its mass:

$$t_{ev} = \frac{c^2 M_0^3}{3K_{ev}} = \left(\frac{c^2 M_0^3}{3}\right)\left(\frac{15360\pi G^2}{\hbar c^6}\right) = \frac{5120\pi G^2 M_0^3}{\hbar c^4} = 8.4109 \times 10^{-17}\left[\frac{M_0}{\text{kg}}\right]^3 \text{ s}$$

Therefore, the evaporation time of a black hole is also proportional to its volume (and the cube of its Schwarzschild radius):

$$r_s = \frac{2GM_0}{c^2}$$

$$V_0 = \frac{4\pi r_s^3}{3} = \frac{32\pi G^3 M_0^3}{3c^6}$$

$$t_{ev} = \frac{5120\pi G^2 M_0^3}{\hbar c^4} = \frac{480 c^2 V_0}{\hbar G} = 6.129\times10^{63}\frac{V_0}{m^3} \, s$$

The time that the black hole takes to dissipate is:

$$t_{ev} = \frac{5120\pi G^2 M_0^3}{\hbar c^4} = \frac{480 c^2 V_0}{\hbar G}$$

where $M_0$ and $V_0$ are the mass and (Schwarzschild) volume of the black hole.

The lower classical quantum limit for mass for this equation is equivalent to the Planck mass, $m_P$.

Hawking radiation evaporation time for a Planck mass quantum black hole:

$$t_{ev} = \frac{5120\pi G^2 m_P^3}{\hbar c^4} = 5120\pi t_P = 5120\pi\sqrt{\frac{\hbar G}{c^5}} = 8.671\times10^{-40} \, s$$

$$t_{ev} = 5120\pi\sqrt{\frac{\hbar G}{c^5}}$$

where $t_P$ is the Planck time.

For a black hole of one solar mass ($M_\odot = 1.98892\times10^{30}$ kg), we get an evaporation time of $2.098\times10^{67}$ years—much longer than the current age of the universe at $(13.799\pm0.021)\times10^9$ years.

$$t_{ev} = \frac{5120\pi G^2 M_\odot^3}{\hbar c^4} = 6.617\times10^{74} \, s$$

But for a black hole of $10^{11}$ kg, the evaporation time is 2.667 billion years. This is why some astronomers are searching for signs of exploding primordial black holes.

However, since the universe contains the cosmic microwave background radiation, in order for the black hole to dissipate, it must have a temperature greater than that of the present-day blackbody radiation of the universe of 2.7 K = $2.3\times10^{-4}$ eV. This implies that $M$ must be less than 0.8% of the mass of the Earth – approximately the mass of the Moon.

Cosmic microwave background radiation universe temperature:

$$T_u = 2.725 \, K$$

Hawking total black hole mass:

$$M_H \leq \frac{\hbar c^3}{8\pi G k_B T_u} \leq 4.503\times10^{22} \, kg$$

$$M_{\mathrm{H}} \leq \frac{\hbar c^3}{8\pi G k_{\mathrm{B}} T_{\mathrm{u}}}$$

$$\frac{M_{\mathrm{H}}}{M_{\oplus}} = 7.539 \times 10^{-3} = 0.754\,\%$$

where $M_{\oplus}$ is the total Earth mass.

In common units,

$$P = 3.56345 \times 10^{32} \left[\frac{\mathrm{kg}}{M}\right]^2 \mathrm{W}$$

$$t_{\mathrm{ev}} = 8.41092 \times 10^{-17} \left[\frac{M_0}{\mathrm{kg}}\right]^3 \mathrm{s} \quad \approx 2.66532 \times 10^{-24} \left[\frac{M_0}{\mathrm{kg}}\right]^3 \mathrm{yr}$$

$$M_0 = 2.28271 \times 10^5 \left[\frac{t_{\mathrm{ev}}}{\mathrm{s}}\right]^{\frac{1}{3}} \mathrm{kg} \quad \approx 7.2 \times 10^7 \left[\frac{t_{\mathrm{ev}}}{\mathrm{yr}}\right]^{\frac{1}{3}} \mathrm{kg}$$

So, for instance, a 1-second-life black hole has a mass of $2.28 \times 10^5$ kg, equivalent to an energy of $2.05 \times 10^{22}$ J that could be released by $5 \times 10^6$ megatons of TNT. The initial power is $6.84 \times 10^{21}$ W.

Black hole evaporation has several significant consequences:

- Black hole evaporation produces a more consistent view of black hole thermodynamics by showing how black holes interact thermally with the rest of the universe.

- Unlike most objects, a black hole's temperature increases as it radiates away mass. The rate of temperature increase is exponential, with the most likely endpoint being the dissolution of the black hole in a violent burst of gamma rays. A complete description of this dissolution requires a model of quantum gravity, however, as it occurs when the black hole approaches Planck mass and Planck radius.

- The simplest models of black hole evaporation lead to the black hole information paradox. The information content of a black hole appears to be lost when it dissipates, as under these models the Hawking radiation is random (it has no relation to the original information). A number of solutions to this problem have been proposed, including suggestions that Hawking radiation is perturbed to contain the missing information, that the Hawking evaporation leaves some form of remnant particle containing the missing information, and that information is allowed to be lost under these conditions.

## Large Extra Dimensions

The previous formulae are only applicable if the laws of gravity are approximately valid all the way down to the Planck scale. In particular, for black holes with masses below the Planck mass (~$10^{-8}$ kg), they result in impossible lifetimes below the Planck time (~$10^{-43}$ s). This is normally seen as an indication that the Planck mass is the lower limit on the mass of a black hole.

In a model with large extra dimensions (10 or 11), the values of Planck constants can be radically different, and the formulae for Hawking radiation have to be modified as well. In particular, the lifetime of a micro black hole with a radius below the scale of the extra dimensions is given by equation 9 in Cheung (2002) and equations 25 and 26 in Carr (2005).

$$\tau \sim \frac{1}{M_*}\left(\frac{M_{BH}}{M_*}\right)^{\frac{n+3}{n+1}}$$

where $M_*$ is the low energy scale, which could be as low as a few TeV, and $n$ is the number of large extra dimensions. This formula is now consistent with black holes as light as a few TeV, with lifetimes on the order of the "new Planck time" ~$10^{-26}$ s.

## In Loop Quantum Gravity

A detailed study of the quantum geometry of a black hole event horizon has been made using loop quantum gravity. Loop-quantization reproduces the result for black hole entropy originally discovered by Bekenstein and Hawking. Further, it led to the computation of quantum gravity corrections to the entropy and radiation of black holes.

Based on the fluctuations of the horizon area, a quantum black hole exhibits deviations from the Hawking spectrum that would be observable were X-rays from Hawking radiation of evaporating primordial black holes to be observed. The quantum effects are centered at a set of discrete and unblended frequencies highly pronounced on top of Hawking radiation spectrum.

## Experimental Observation

Under experimentally achievable conditions for gravitational systems this effect is too small to be observed directly. However, in September 2010 an experimental set-up created a laboratory "white hole event horizon" that the experimenters claimed was shown to radiate an optical analog to Hawking radiation, although its status as a genuine confirmation remains in doubt. Some scientists predict that Hawking radiation could be studied by analogy using sonic black holes, in which sound perturbations are analogous to light in a gravitational black hole and the flow of an approximately perfect fluid is analogous to gravity.

# References

- Carroll, sean (2003). Spacetime and geometry: an introduction to general relativity. Isbn 0-8053-8732-3

- Black-holes: sciencealert.com, Retrieved , 24 March , 2019

- Bennett, jay (april 10, 2019). "astronomers capture first-ever image of a supermassive black hole". Smithsonian.com. Smithsonian institute. Retrieved april 15, 2019

- Topics_blackholes_event: physicsoftheuniverse.com, Retrieved 12 April, 2019

- Henderson, mark (september 9, 2008). "stephen hawkings 50 bet on the world the universe and the god particle". The times. London. Retrieved may 4, 2010

- Singularity: universetoday.com, Retrieved 31 July, 2019

- Teo, edward (2003). "spherical photon orbits around a kerr black hole" (pdf). General relativity and gravitation. 35 (11): 1909–1926. Bibcode:2003gregr..35.1909t. Doi:10.1023/a:1026286607562. Issn 0001-7701

# PERMISSIONS

# INDEX

**A**

Asteroid Belt, 2, 11-12, 140, 145
Asteroids, 6-7, 10-12, 82, 138-140, 145
Astronomical Distances, 2
Astronomical Unit, 5, 65, 75-81, 159
Atmospheric Effects, 3, 34

**B**

Baryons, 49, 97, 99-100, 118, 122, 124
Beryllium, 51
Big Bang Model, 32, 51, 95-97
Blackbody Spectrum, 17, 31

**C**

Cassini Spacecraft, 6, 10
Celestial Sphere, 68-70, 74, 151
Cepheid Variables, 3, 21, 170, 172, 177
Cloud Cover, 9
Computational Astrophysics, 42, 46
Cosmic Footprints, 7
Cosmic Inflation, 93, 96, 99, 102-104, 110
Cosmology, 1-2, 30, 39, 47-48, 50, 94-96, 98, 104-106, 109, 114, 117, 127

**D**

Dark Energy, 33, 49, 65, 93, 104, 117, 122, 126, 132-134, 176
Deuterium, 31, 51, 136, 148
Doppler Shift, 4, 40, 96
Dust Belt, 15, 163
Dwarf Planets, 6-7, 12

**E**

Earth Mass, 82, 85-87, 90-91, 219
Electromagnetic Radiation, 10, 36, 94, 117, 125, 187, 198, 209
Energy Radiation, 53, 97

**G**

Galactic Cosmic Rays, 9, 27

Galactic Plane, 74, 105, 167
Gamma Rays, 30, 35-36, 49, 125, 130-131, 200, 219
Gaussian Gravitational Constant, 76, 79
Geographical Latitude, 67-68, 71
Globular Cluster, 25, 61, 167, 171, 174
Gravitational Acceleration, 45-46, 198
Gravitational Potential, 51, 61-62, 67, 108, 120, 122
Gravity Waves, 36, 38

**H**

Heliosphere, 55, 126
Helium, 1, 9, 13, 18, 21, 27, 31-32, 38, 41, 51-52, 82, 84, 91, 93, 100, 124, 139, 144, 150, 157, 176, 181, 187
Hipparcos Satellite, 3, 149
Hubble Law, 4, 30
Hydrogen Fusion, 51, 136

**I**

Infrared Wavelengths, 9, 34, 36, 154
Interstellar Gas, 3, 135, 179, 182, 191, 196
Interstellar Medium, 21, 25, 27-28, 43, 58, 107, 112, 144, 167, 182, 186-188, 190, 193-196
Interstellar Molecules, 25, 53

**J**

Jupiter Mass, 82-84

**K**

Kinetic Energy, 27, 62, 135, 191-192
Kuiper Belt, 6-7, 11-12, 15, 80, 141

**L**

Laws Of Motion, 1, 38, 44
Lithium, 1, 51, 124
Luminosity, 2-3, 17-21, 26, 29, 65, 110-116, 137, 146, 154, 157-158, 164-167, 169-172, 174, 177, 216
Lunar Exploration, 8

**M**

Magellanic Clouds, 30, 168

Magnetic Field, 9-10, 27, 56-58, 102, 155, 178

Mass Spectrometer, 37

Mercury, 4-10, 14, 80, 85

Meridians, 67

Meteoroids, 6, 12-13

Milky Way Galaxy, 1, 24-27, 30, 63, 66, 74, 105, 149, 161, 180-181

**N**

Nebulae, 39-40, 81, 92, 136, 181-185, 188, 194, 196

Neptune, 4-7, 10-12, 15, 40, 84, 86, 139, 141

Neutrino, 33, 47-50, 52, 125, 127-129, 131, 214

Nuclear Fusion, 40-41, 60, 82, 186

**O**

Oort Cloud, 6, 11, 80, 141

**P**

Partial Differential Equations, 44

**R**

Radiated Power, 2, 111, 210

Radio Emission, 10, 178, 186, 203

Relative Mass, 81, 84

Rubidium, 13

**S**

Solar Mass, 20, 78-79, 81-82, 85-86, 147, 150, 159, 163, 210, 213-216, 218

Solar System, 1-2, 4-6, 9-11, 13-14, 16, 24, 44, 51-52, 55, 68, 74-83, 89, 118, 130, 132, 139-141, 145, 148, 150, 161, 178, 180, 190

Solar Wind, 9-10, 56, 82, 147, 190

Spectral Analysis, 13, 40, 155

Stellar Atmospheres, 43, 53, 155

Stellar Distances, 2-3, 76, 148-149

Stellar Radii, 164

Stratosphere, 2, 13, 37

Supernova Explosions, 3, 27, 58, 162, 183, 193

Synchrotron Radiation, 10, 27, 38

**T**

Terrestrial Planets, 6-7, 14, 85, 135, 138-139

Thorium, 13

Triangulation, 2

**U**

Uranium, 13, 51

Uranus, 4, 6-7, 9-10, 15, 40, 84, 86, 139

**V**

Venus, 4, 6-10, 15, 80-81, 85

**Z**

Zero Meridian, 69